SNIPER TRAINING

I0118722

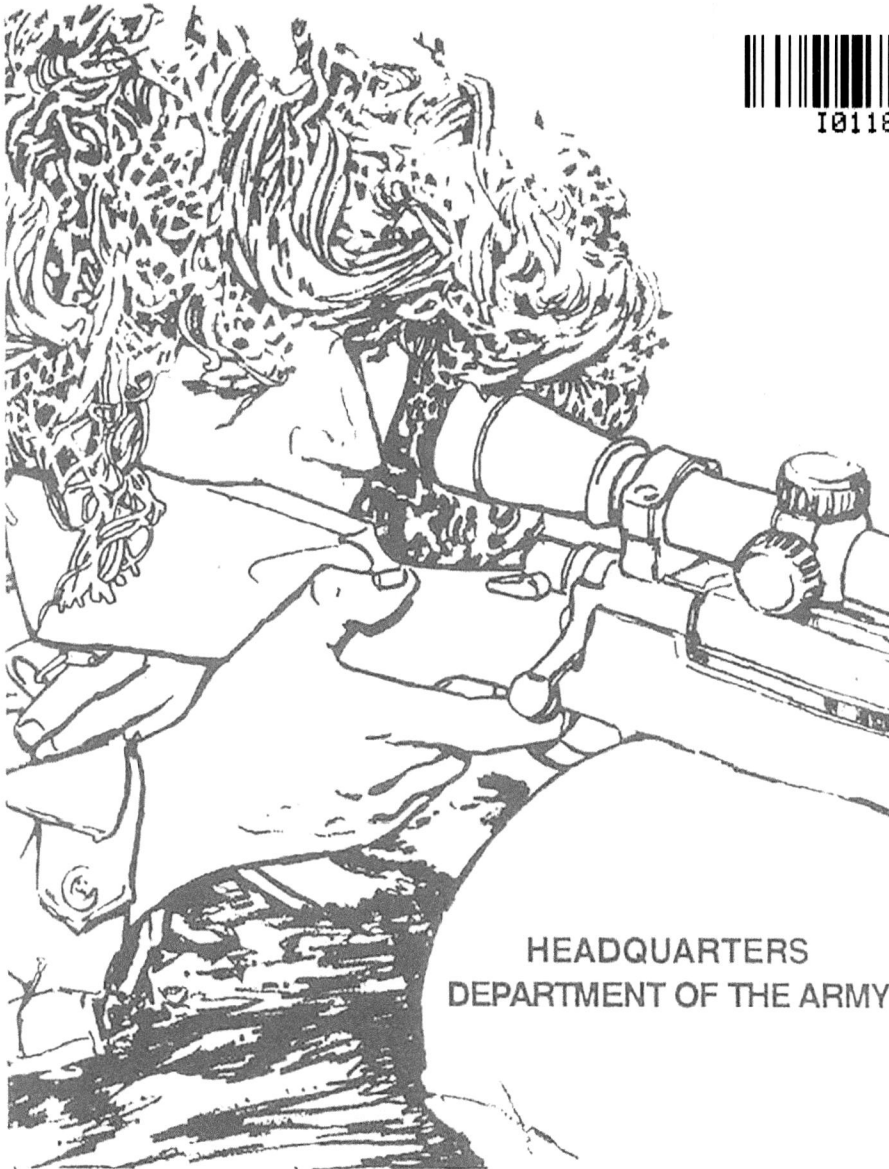

HEADQUARTERS
DEPARTMENT OF THE ARMY

DISTRIBUTION RESTRICTION—
Approved for public release; distribution is unlimited.

*FM 23-10
17 August 1994

FIELD MANUAL
No. 23-10

HEADQUARTERS
DEPARTMENT OF THE ARMY
Washington, DC, 17 August 1994

SNIPER TRAINING

CONTENTS

DISTRIBUTION RESTRICTION: Approved for public release; distribution is unlimited.

*This publication supersedes TC 23-14, 14 June 1989.

PREFACE

This field manual provides information needed to train and equip snipers and to aid them in their missions and operations. It is intended for use by commanders, staffs, trainers, snipers, and soldiers at training posts, Army schools, and units.

This manual is organized as a reference for snipers and leads the trainer through the material needed to conduct sniper training. Subjects include equipment, weapon capabilities, fundamentals of marksmanship and ballistics, field skills, mission planning, and skill sustainment. The left-handed firer can become a sniper, but all material in this book is referenced to the right-handed firer.

The proponent for this publication is Headquarters, United States Army Infantry School. Send comments and recommendations on DA Form 2028 (Recommended Changes to Publications and Blank Forms) directly to the Commandant, United States Army Infantry School, ATTN: ATSH-IN-S3, Fort Benning, GA 31905-5596.

This publication complies with the following international agreements:

STANAG 2020	Operational Situation Report
STANAG 2022	Intelligence Report
STANAG 2084	Handling and Reporting of Captured Enemy Equipment and Documents
STANAG 2103	Reporting Nuclear Detonations, Radioactive Fallout and Biological and Chemical Attacks, and Predicting Associated Hazards
STANAG 2934	Artillery Procedures—AARTY-l
STANAG 3204	Aeromedical Evacuation
STANAG 6004	Meaconing, Intrusion, Jamming, and Interference Report

> Unless this publication states otherwise, masculine nouns and pronouns do not refer exclusively to men.

CHAPTER 1
INTRODUCTION

The sniper has special abilities, training and equipment. His job is to deliver discriminatory highly accurate rifle fire against enemy targets, which cannot be engaged successfully by the rifleman because of range, size, location, fleeting nature, or visibility. Sniping requires the development of basic infantry skills to a high degree of perfection. A sniper's training incorporates a wide variety of subjects designed to increase his value as a force multiplier and to ensure his survival on the battlefield. The art of sniping requires learning and repetitiously practicing these skills until mastered. A sniper must be highly trained in long-range rifle marksmanship and field craft skills to ensure maximum effective engagements with minimum risk.

1-1. MISSION

The primary mission of a sniper in combat is to support combat operations by delivering precise long-range fire on selected targets. By this, the sniper creates casualties among enemy troops, slows enemy movement, frightens enemy soldiers, lowers morale, and adds confusion to their operations. The secondary mission of the sniper is collecting and reporting battlefield information.

a. A well-trained sniper, combined with the inherent accuracy of his rifle and ammunition, is a versatile supporting arm available to an infantry commander. The importance of the sniper cannot be measured simply by the number of casualties he inflicts upon the enemy. Realization of the sniper's presence instills fear in enemy troop elements and influences their decisions and actions. A sniper enhances a unit's firepower and augments the varied means for destruction and harassment of the enemy. Whether a sniper is organic or attached, he will provide that unit with extra supporting fire. The sniper's role is unique in that it is the sole means by which a unit can engage point targets at distances beyond the effective range of the M16 rifle. This role becomes more significant when the target is entrenched or positioned among civilians, or during riot control missions. The fires of automatic weapons in such operations can result in the wounding or killing of noncombatants.

b. Snipers are employed in all levels of conflict. This includes conventional offensive and defensive combat in which precision fire is delivered at long ranges. It also includes combat patrols, ambushes, countersniper operations, forward observation elements, military operations in urbanized terrain, and retrograde operations in which snipers are part of forces left in contact or as stay-behind forces.

1-2. ORGANIZATION

In light infantry divisions, the sniper element is composed of six battalion personnel organized into three 2-man teams. The commander designates missions and priorities of targets for the team and may attach or place the team under the operational control of a company or platoon. They may perform dual missions, depending on the need. In the mechanized infantry battalions, the sniper element is composed of two riflemen (one team) located in a rifle squad. In some specialized units, snipers may be organized according to the needs of the tactical situation.

a. Sniper teams should be centrally controlled by the commander or the sniper employment officer. The SEO is responsible for the command and control of snipers assigned to the unit. In light infantry units, the SEO will be the reconnaissance platoon leader or the platoon sergeant. In heavy or mechanized units, the SEO may be the company commander or the executive officer. The duties and responsibilities of the SEO areas follows:

(1) To advise the unit commander on the employment of snipers.
(2) To issue orders to the team leader.
(3) To assign missions and types of employment.
(4) To coordinate between the sniper team and unit commander.
(5) To brief the unit commander and team leaders.
(6) To debrief the unit commander and team leaders.
(7) To train the teams.

b. Snipers work and train in 2-man teams. One sniper's primary duty is that of the sniper and team leader while the other sniper serves as the observer. The sniper team leader is responsible for the day-to-day activities of the sniper team. His responsibilities areas follows:

(1) To assume the responsibilities of the SEO that pertain to the team in the SEO'S absence.

(2) To train the team.

(3) To issue necessary orders to the team.

(4) To prepare for missions.

(5) To control the team during missions.

c. The sniper's weapon is the sniper weapon system. The observer has the M16 rifle and an M203, which gives the team greater suppressive fire and protection. Night capability is enhanced by using night observation devices.

1-3. PERSONNEL SELECTION CRITERIA

Candidates for sniper training require careful screening. Commanders must screen the individual's records for potential aptitude as a sniper. The rigorous training program and the increased personal risk in combat require high motivation and the ability to learn a variety of skills. Aspiring snipers must have an excellent personal record.

a. The basic guidelines used to screen sniper candidates are as follows:

(1) *Marksmanship.* The sniper trainee must be an expert marksman. Repeated annual qualification as expert is necessary. Successful participation in the annual competition-in-arms program and an extensive hunting background also indicate good sniper potential.

(2) *Physical condition.* The sniper, often employed in extended operations with little sleep, food, or water, must be in outstanding physical condition. Good health means better reflexes, better muscular control, and greater stamina. The self-confidence and control that come from athletics, especially team sports, are definite assets to a sniper trainee.

(3) *Vision.* Eyesight is the sniper's prime tool. Therefore, a sniper must have 20/20 vision or vision correctable to 20/20. However, wearing glasses could become a liability if glasses are lost or damaged. Color blindness is also considered a liability to the sniper, due to his inability to detect concealed targets that blend in with the natural surroundings.

(4) *Smoking.* The sniper should not be a smoker or use smokeless tobacco. Smoke or an unsuppressed smoker's cough can betray the sniper's position. Even though a sniper may not smoke or use smokeless tobacco on a mission, his refrainment may cause nervousness and irritation, which lowers his efficiency.

(5) *Mental condition.* When commanders screen sniper candidates, they should look for traits that indicate the candidate has the right qualities to be a sniper. The commander must determine if the candidate will pull the trigger at the right time and place. Some traits to look for

are reliability, initiative, loyalty, discipline, and emotional stability. A psychological evaluation of the candidate can aid the commander in the selection process.

(6) *Intelligence.* A sniper's duties require a wide variety of skills. He must learn the following:

- Ballistics.
- Ammunition types and capabilities.
- Adjustment of optical devices.
- Radio operation and procedures.
- Observation and adjustment of mortar and artillery fire.
- Land navigation skills.
- Military intelligence collecting and reporting.
- Identification of threat uniforms and equipment.

b. In sniper team operations involving prolonged independent employment, the sniper must be self-reliant, display good judgment and common sense. This requires two other important qualifications: emotional balance and field craft.

(1) *Emotional balance.* The sniper must be able to calmly and deliberately kill targets that may not pose an immediate threat to him. It is much easier to kill in self-defense or in the defense of others than it is to kill without apparent provocation. The sniper must not be susceptible to emotions such as anxiety or remorse. Candidates whose motivation toward sniper training rests mainly in the desire for prestige may not be capable of the cold rationality that the sniper's job requires.

(2) *Field craft.* The sniper must be familiar with and comfortable in a field environment. An extensive background in the outdoors and knowledge of natural occurrences in the outdoors will assist the sniper in many of his tasks. Individuals with such a background will often have great potential as a sniper.

c. Commander involvement in personnel selection is critical. To ensure his candidate's successful completion of sniper training and contribution of his talents to his unit's mission, the commander ensures that the sniper candidate meets the following prerequisites before attending the US Army Sniper School:

- Male.
- PFC to SFC (waiverable for MSG and above).
- Active duty or ARNG and USAR.
- Good performance record.

- No history of alcohol or drug abuse.
- A volunteer (with commander recommendation).
- Vision of 20/20 or correctable to 20/20.
- No record of disciplinary action.
- Expert marksman with M16A1 or M16A2 rifle.
- Minimum of one-year retrainability.
- Career management field 11.
- Pass APFT (70 percent, each event).

1-4. SNIPER AND OBSERVER RESPONSIBILITIES

Each member of the sniper team has specific responsibilities. Only through repeated practice can the team begin to function properly. Responsibilities of team members areas follows:

a. The sniper—
- Builds a steady, comfortable position.
- Locates and identifies the designated target.
- Estimates the range to the target.
- Dials in the proper elevation and windage to engage the target.
- Notifies the observer of readiness to fire.
- Takes aim at the designated target.
- Controls breathing at natural respiratory pause.
- Executes proper trigger control.
- Follows through.
- Makes an accurate and timely shot call.
- Prepares to fire subsequent shots, if necessary.

b. The observer—
- Properly positions himself.
- Selects an appropriate target.
- Assists in range estimation.
- Calculates the effect of existing weather conditions on ballistics.
- Reports sight adjustment data to the sniper.
- Uses the M49 observation telescope for shot observation.
- Critiques performance.

1-5. TEAM FIRING TECHNIQUES

A sniper team must be able to move and survive in a combat environment. The sniper team's mission is to deliver precision fire. This calls for a coordinated team effort. Together, the sniper and observer—

- Determine the effects of weather on ballistics.
- Calculate the range to the target.
- Make necessary sight changes.
- Observe bullet impact.
- Critique performance before any subsequent shots.

CHAPTER 2

EQUIPMENT

This chapter describes the equipment necessary for the sniper to effectively peform his mission. The sniper carries only what is essential to successfully complete his mission. He requires a durable rifle with the capability of long-range precision fire. The current US Army sniper weapon system is the M24. (See Appendix B for the M21 sniper weapon system.)

Section I
M24 SNIPER WEAPON SYSTEM

The M24 sniper weapon system is a 7.62-mm, bolt-action, six-shot repeating rifle (one round in the chamber and five rounds in the magazine). It is designed for use with either the M3A telescope (day optic sight) (usually called the M3A *scope)* or the metallic iron sights. The sniper must know the M24's components, and the procedures required to operate them (Figure 2-1, page 2-2). The deployment kit is a repair/ maintenance kit with tools and repair parts for the operator to perform operator level maintenance (Figure 2-2, page 2-3.)

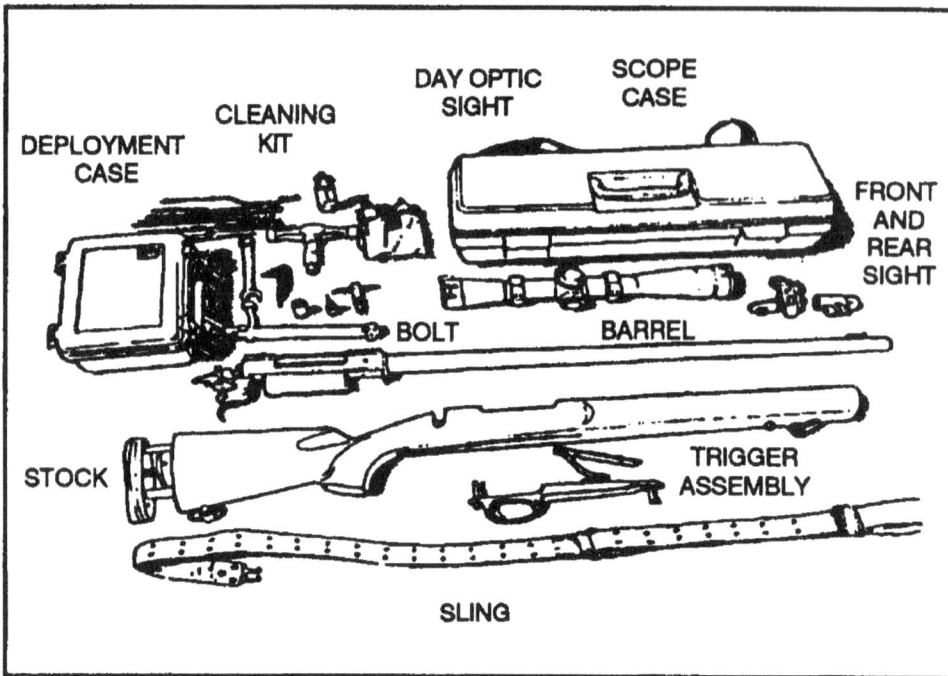

Figure 2-1. M24 sniper weapon system.

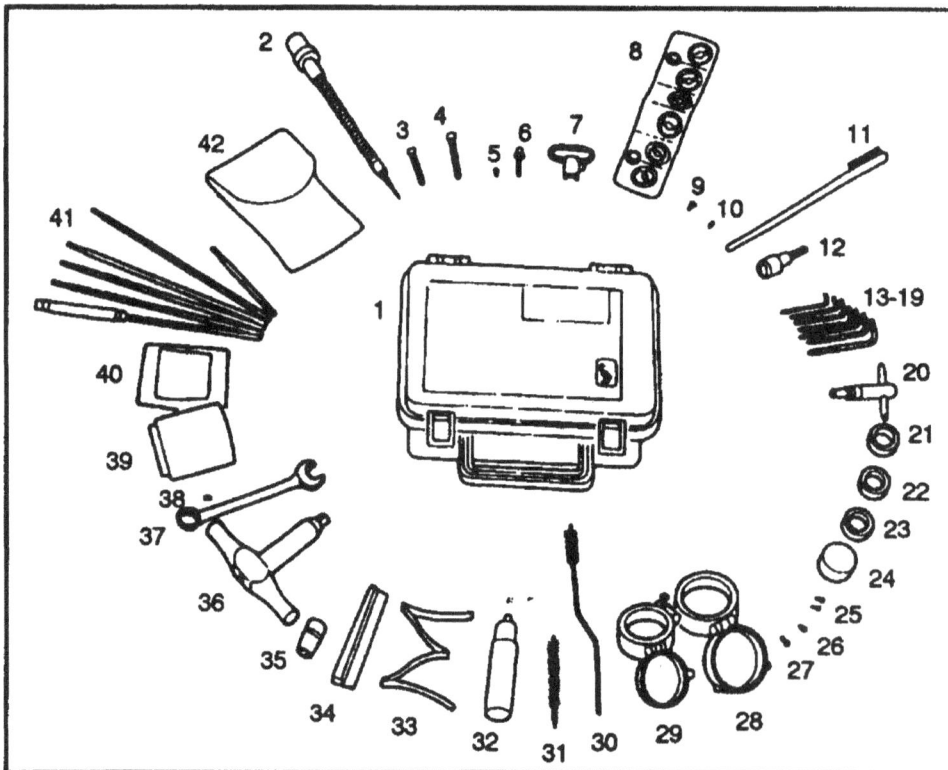

Figure 2-2. The deployment kit.

1 DEPLOYMENT CASE

2 FIRING PIN ASSEMBLY

3 FRONT GUARD SCREW

4 REAR GUARD SCREWS

5 FRONT SIGHT BASE SCREW

6 SWIVEL SCREW

7 SWIVEL, SLING

8 FRONT SIGHT INSERT KIT

9 REAR SIGHT BASE SCREW

10 TRIGGER PULL ADJUSTMENT SCREW

11 BRUSH, CLEANING SMALL

12 SOCKET WRENCH ATTACHMENT 3/8-INCH DRIVE HEX BIT 5/32-INCHES

13 050-INCH KEY, SOCKET HEAD SCREW

14 1/16-INCH KEY, SOCKET

15 5/64-INCH KEY, SOCKET HEAD SCREW

16 3/32-INCH KEY, SOCKET HEAD SCREW

17 7/64-INCH KEY, SOCKET HEAD SCREW

18 1/8-INCH KEY, SOCKET HEAD SCREW

19 5/32-INCH KEY, SOCKET HEAD SCREW

20 T-HANDLE COMBINATION WRENCH

21 DAY OPTIC SIGHT WINDAGE DIAL WITH SCREWS

22 DAY OPTICT SIGHT ELEVATION DIAL WITH SCEWS

23 DAY OPTIC FOCUS DIAL WITH SCREWS

24 DAY OPTIC SIGHT ADJUSTMENT DIAL DUST COVER

25 DAY OPTIC SIGHT RING SCREWS

26 DAY OPTIC SIGHT BASE SCREWS

27 DAY OPTIC SIGHT BASE REAR

28 DAY OPTIC SIGHT DUST COVER, FRONT

29 DAY OPTIC SIGHT DUST COVER, REAR

30 BRUSH, CHAMBER

31 BRUSH, BORE

32 OIL BOTTLE

33 MAGAZINE SPRING

34 MAGAZINE FOLLOWER

35 SOCKET, SOCKET WRENCH HEAD SCREW 1/2-INCH

36 T-HANDLE TORQUE WRENCH

37 WRENCH, BOX AND OPEN 1/2-INCH

38 REAR SIGHT BASE PLUG SCREW

39 DAY OPTIC SIGHT SUNSHADE

40 SWABS, CLEANING, SMALL ARMS

41 CLEANING ROD KIT

42 LENS CLEANING KIT

Figure 2-2. The deployment kit (continued).

2-1. OPERATIONS AND FUNCTIONS

To operate the M24 sniper weapon system, the sniper must know the information and instructions pertaining to the safety, bolt assembly, trigger assembly, and stock adjustment.

a. **Safety.** The safely is located on the right rear side of the receiver. When properly engaged, the safety provides protection against accidental discharge in normal usage.

(1) To engage the safety, place it in the "S" position (Figure 2-3).

(2) Always place the safety in the "S" position before handling, loading, or unloading the weapon.

(3) When the weapon is ready to be fired, place the safety in the "F" position (Figure 2-3).

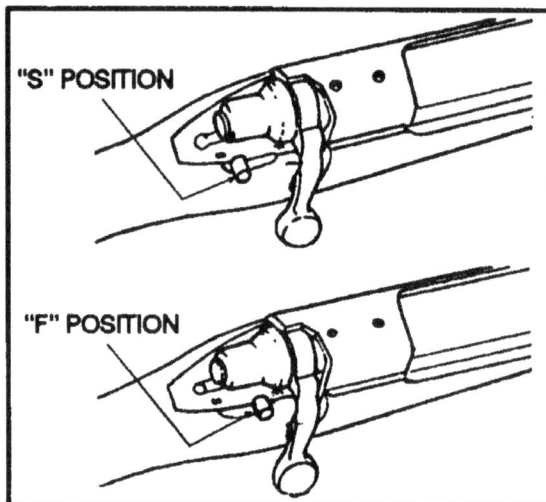

b. **Bolt Assembly.** The bolt assembly locks the cartridge into the chamber and extracts the cartridge from the chamber.

(1) To remove the bolt from the receiver, release the internal magazine, place the safety in the "S" position, raise the bolt handle, and pull it back until it stops. Then push the bolt stop release (Figure 2-4) and pull the bolt from the receiver.

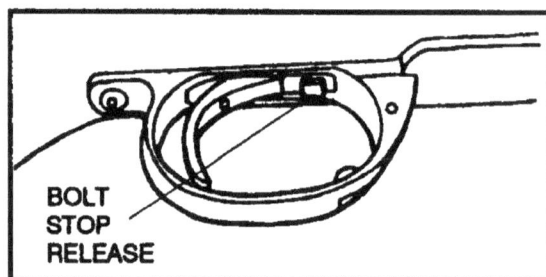

(2) To replace the bolt, ensure the safety is in the "S" position, align the lugs on the bolt assembly with the receiver (Figure 2-5), slide the bolt all the way into the receiver, and then push the bolt handle down.

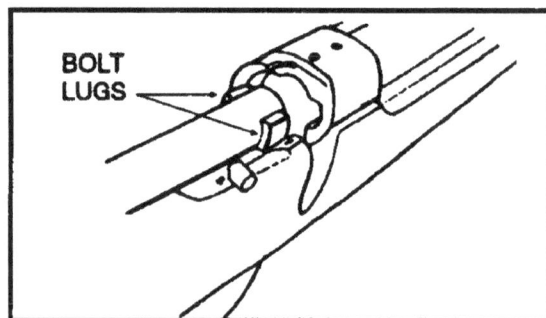

Figure 2-3. Safety.

Figure 2-4. Bolt stop release.

Figure 2-5. Bolt alignment.

> **WARNING**
> **NEVER REMOVE THE TRIGGER MECHANISM, OR MAKE ADJUSTMENTS TO THE TRIGGER ASSEMBLY, EXCEPT FOR THE TRIGGER PULL FORCE ADJUSTMENT.**

c. **Trigger Assembly.** Pulling the trigger fires the rifle when the safety is in the "F" position. The operator may adjust the trigger pull force from a minimum of 2 pounds to a maximum of 8 pounds. This is done using the l/16-inch socket head screw key provided in the deployment kit. Turning the trigger adjustment screw (Figure 2-6) clockwise increases the force needed to pull the trigger. Turning it counterclockwise decreases the force needed. This is the only trigger adjustment the sniper should make.

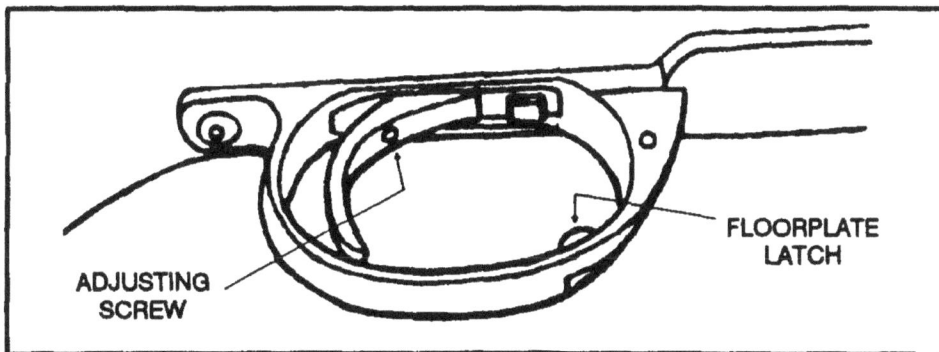

Figure 2-6. Trigger adjustment.

d. **Stock Adjustment.** The M24's stock has an adjustable butt plate to accommodate the length of pull. The stock adjustment (Figure 2-7) consists of a thin wheel and a thick wheel. The thick wheel adjusts the shoulder stock. The thin wheel locks the shoulder stock.

(1) Turn the thick wheel clockwise to *lengthen* the stock.

(2) Turn the thick wheel counterclockwise to *shorten* the stock.

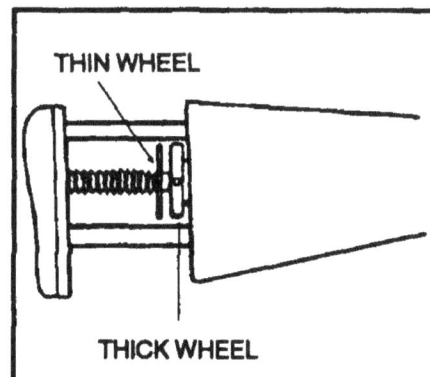

Figure 2-7. Stock adjustment.

(3) To lock the shoulder stock into position, turn the thin wheel clockwise against the thick wheel.

(4) To unlock the shoulder stock, turn the thin wheel counter-clockwise away from the thick wheel.

e. **Sling Adjustment** The sling helps hold the weapon steady without muscular effort. The more the muscles are used the harder it is to hold the weapon steady. The sling tends to bind the parts of the body used in aiming into a rigid bone brace, requiring less effort than would be necessary if no sling were used. When properly adjusted, the sling permits part of the recoil of the rifle to reabsorbed by the nonfiring arm and hand, removing recoil from the firing shoulder.

(1) The sling consists of two different lengths of leather straps joined together by a metal D ring (Figure 2-8). The longer strap is connected to the sling swivel on the rear stud on the forearm of the rifle. The shorter strap is attached to the sling swivel on the buttstock of the rifle. There are two leather loops on the long strap known as keepers. The keepers are used to adjust the tension on the sling. The frogs are hooks that are used to adjust the length of the sling.

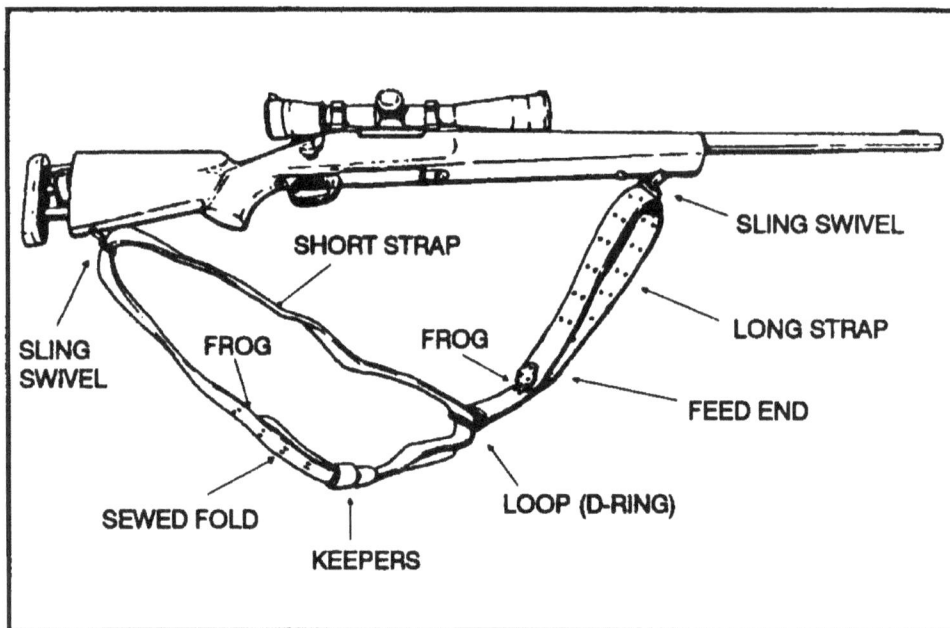

Figure 2-8. Leather Sling.

(2) To adjust the sling, the sniper disconnects the sling from the buttstock swivel. Then, he adjusts the length of the metal D ring that joins

the two halves of the sling. He then makes sure it is even with the comb of the stock when attaching the sling to the front swivel (Figure 2-9).

Figure 2-9. Sling adjustment.

2-10. Adjusting the length of the sling.

(3) The sniper adjusts the length of the sling by placing the frog on the long strap of the sling in the 4th to the 7th set of adjustment holes on the rounded end of the long strap that goes through the sling swivel on the forearm (Figure 2-10).

(4) After adjusting the length, the sniper places the weapon on his firing hip and supports the

weapon with his firing arm. The sniper turns the sling away from him 90 degrees and inserts his nonfiring arm.

(5) The sniper slides the loop in the large section of the sling up the nonfiring arm until it is just below the armpit (Figure 2-11). He then slides both leather keepers down the sling until they bind the loop snugly round the nonfiring arm.

Figure 2-11. Placing the sling around the nonfiring arm.

Figure 2-12. Proper placement of the sling.

(6) The sniper moves his nonfiring hand from the outside of the sling to the inside of the sling between the rifle and the sling. The sniper then grasps the forearm of the weapon, just behind the sling swivel with his nonfiring hand. He forces it outward and away from his body with the nonfiring hand (Figure 2-12).

(7) The sniper pulls the butt of the

weapon into the pocket of his shoulder with the firing hand. He then grasps the weapon at the small of the stock and begins the aiming process.

2-2. INSPECTION

The sniper performs PMCS on the M24 SWS. Deficiencies that cannot be repaired by the sniper requires manufacturer repair. He must refer to TM 9-1005-306-10 that is furnished with each weapon system. The sniper must know this technical manual. He should cheek the following areas when inspecting the M24:

a. Check the appearance and completeness of all parts.

b. Check the bolt to ensure it locks, unlocks, and moves smoothly.

c. Check the safety to ensure it can be positively placed into the "S" and "F" positions easily without being too hard or moving too freely.

d. Check the trigger to ensure the weapon will not fire when the safety is in the "S" position, and that it has a smooth, crisp trigger pull when the safety is in the "F" position.

e. Check the trigger guard screws (rear of trigger guard and front of internal magazine) for proper torque (65 inch-pounds).

f. Check the scope mounting ring nuts for proper torque (65 inch-pounds).

g. Check the stock for any cracks, splits, or any contact it may have with the barrel.

h. Inspect the scope for obstructions such as dirt, dust, moisture, or loose or damaged lenses.

2-3. CARE AND MAINTENANCE

Maintenance is any measure taken to keep the M24 SWS in top operating condition. It includes inspection, repair, cleaning and lubrication- Inspection reveals the need for repair, cleaning, or lubrication. It also reveals any damages or defects. When sheltered in garrison and infrequently used, the M24 SWS must be inspected often to detect dirt, moisture, and signs of corrosion, and it must be cleaned accordingly. The M24 SWS that is in use and subject to the elements, however, requires no inspection for cleanliness, since the fact of its use and exposure is evidence that it requires repeated cleaning and lubrication.

a. **M24 SWS Maintenance.** The following materials are required for cleaning and maintaining the M24 SWS:

- One-piece plastic-coated .30 caliber cleaning rod with jag (36 inches).
- Bronze bristle bore brushes (.30 and .45 calibers).
- Cleaning patches (small and large sizes).

- Carbon cleaner.
- Copper cleaner.
- Rust prevention.
- Cleaner, lubricant, preservative.
- Rifle grease.
- Bore guide (long action).
- Swabs.
- Pipe cleaners.
- Medicine dropper.
- Shaving brush.
- Pistol cleaning rod.
- Rags.
- Camel's-hair brush.
- Lens tissue.
- Lens cleaning fluid (denatured or isopropyl alcohol).

b. **M24 SWS Disassembly.** The M24 SWS will be disassembled only when necessary, not for daily cleaning. For example, when removing an obstruction from the SWS that is stuck between the stock and the barrel. When disassembly is required, the recommended procedure is as follows:

- Place the weapon so that is it pointing in a safe direction.
- Ensure the safety is in the "S" position.
- Remove the bolt assembly.
- Loosen the mounting ring nuts on the telescope and remove the telescope.
- Remove the action screws.
- Lift the stock from the barrel assembly.
- For further disassembly, refer to TM 9-1005-306-10.

c. **M24 SWS Cleaning Procedures.** The M24 SWS must always be cleaned *before* and *after firing.*

(1) The SWS must always be cleaned *before firing.* Firing a weapon with a dirty bore or chamber will multiply and speed up any corrosive action. Oil in the bore and chamber of a SWS will cause pressures to vary and first-round accuracy will suffer. Clean and dry the bore and chamber before departure on a mission and use extreme care to keep the SWS clean and dry en route to the objective area. Firing a SWS with oil or moisture in the bore will cause smoke that can disclose the firing position.

(2) The SWS must be cleaned *after firing* since firing produces deposits of primer fouling, powder ashes, carbon, and metal fouling. Although ammunition has a noncorrosive primer that makes cleaning easier, the primer residue can still cause rust if not removed. Firing leaves two major types of fouling that require different solvents to remove *carbon* fouling and *copper* jacket fouling. The SWS must be cleaned within a reasonable time after firing. Use common sense when cleaning between rounds of firing. Repeated firing will not injure the weapon if it is properly cleaned before the first round is fired.

(3) Lay the SWS on a table or other flat surface with the muzzle away from the body and the sling down. Make sure not to strike the muzzle or telescopic sight on the table. The cleaning cradle is ideal for holding the SWS.

(4) Always clean the bore from the chamber toward the muzzle, attempting to keep the muzzle lower than the chamber to prevent the bore cleaner from running into the receiver or firing mechanism. Be careful not to get any type of fluid between the stock and receiver. If fluid does collect between the stock and receiver, the receiver will slide on the bedding every time the SWS recoils, thereby decreasing accuracy and increasing wear and tear on the receiver and bedding material.

(5) Always use a bore guide to keep the cleaning rod centered in the bore during the cleaning process.

(6) Push several patches saturated with carbon cleaner through the barrel to loosen the powder fouling and begin the solvent action on the copper jacket fouling.

(7) Saturate the bronze bristle brush (NEVER USE STAINLESS STEEL BORE BRUSHES-THEY WILL SCRATCH THE BARREL) with carbon cleaner (shake the bottle regularly to keep the ingredients mixed) using the medicine dropper to prevent contamination of the carbon cleaner. Run the bore brush through at least 20 times. Make sure the bore brush passes completely through the barrel before reversing its direction; otherwise, the bristles will break off.

(8) Use a pistol cleaning rod and a .45 caliber bronze bristle bore brush, clean the chamber by rotating the patch-wrapped brush 8 to 10 times. DO NOT scrub the brush in and out of the chamber.

(9) Push several patches saturated with carbon cleaner through the bore to push out the loosened powder fouling.

(10) Continue using the bore brush and patches with carbon cleaner until the patches have no traces of black/ gray powder fouling and are green/ blue. This indicates that the powder fouling has been removed and only copper fouling remains. Remove the carbon cleaner from the barrel

with several clean patches. This is important since solvents should never be mixed in the barrel.

(11) Push several patches saturated with copper cleaner through the bore, using a scrubbing motion to work the solvent into the copper. Let the solvent work for 10 to 15 minutes (NEVER LEAVE THE COPPER CLEANER IN THE BARREL FOR MORE THAN 30 MINUTES).

(12) While waiting, scrub the bolt with the toothbrush moistened with carbon cleaner and wipe down the remainder of the weapon with a cloth.

(13) Push several patches saturated with copper cleaner through the barrel. The patches will appear dark blue at first, indicating the amount of copper fouling removed. Continue this process until the saturated patches have no traces of blue/ green. If the patches continue to come out dark blue after several treatments with copper cleaner, use the bronze brush saturated with copper cleaner to increase the scrubbing action. Be sure to clean the bronze brush thoroughly afterwards with hot running water (quick scrub cleaner/ degreaser is preferred) as the copper cleaner acts upon its bristles as well.

(14) When the barrel is clean, dry it with several tight fitting patches. Also, dry the chamber using the .45 caliber bronze bristle bore brush with a patch wrapped around it.

(15) Run a patch saturated with rust prevention (*not* CLP) down the barrel and chamber if the weapon is to be stored for any length of time. Stainless steel barrels are not immune from corrosion. Be sure to remove the preservative by running dry patches through the bore and chamber before firing.

(16) Place a small amount of rifle grease on the rear surfaces of the bolt lugs. This will prevent galling of the metal surfaces.

(17) Wipe down the exterior of the weapon (if it is not covered with camouflage paint) with a CLP-saturated cloth to protect it during storage.

d. **Barrel Break-in Procedure.** To increase barrel life, accuracy, and reduce cleaning requirement the following barrel break-in procedure must be used. This procedure is best accomplished when the SWS is new or newly rebarreled. The break-in period is accomplished by polishing the barrel surface under heat and pressure. This procedure should only be done by qualified personnel. The barrel must be cleaned of all fouling, both powder and copper. The barrel is dried, and one round is fired. The barrel is then cleaned again using carbon cleaner and then copper cleaner. The barrel must be cleaned again, and another round is fired. The procedure must be repeated for a total of 10 rounds. After the 10th round the SWS is then tested for groups by firing three-round shot groups, with a complete barrel cleaning between shot groups for a total of five shot groups (15 rounds total).

The barrel is now broken in, and will provide superior accuracy and a longer usable barrel life. Additionally, the barrel will be easier to clean because the surface is smoother. Again the barrel should be cleaned at least every 50 rounds to increase the barrel life.

e. **Storage.** The M24 SWS should be stored (Figure 2-13) using the following procedures:

- Clear the SWS, close the bolt, and squeeze the trigger.
- Open the lens caps to prevent gathering of moisture.
- Hang the weapon upside down by the rear sling swivel.
- Place all other items in the system case.
- Transport the weapon in the system case during nontactical situations.
- Protect the weapon at all times during tactical movement.

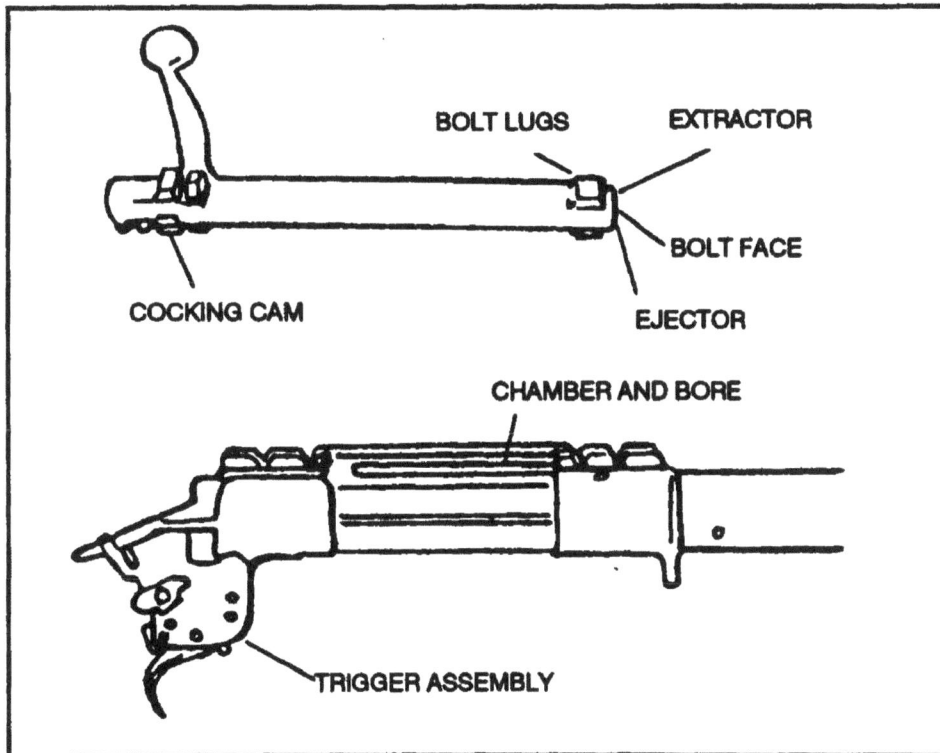

Figure 2-13. Maintenance for storing or using.

NOTE: Rod clean swabs through the bore before firing. This procedure ensures first-round accuracy and reduces the signature.

f. **Cold Climates.** In temperatures below freezing, the SWS must be kept free of moisture and heavy oil, both of which will freeze, causing the working parts to freeze or operate sluggishly. The SWS should be stored in a room with the temperature equal to the outside temperature. When the SWS is taken into a warm area, condensation occurs, thus requiring a thorough cleaning and drying before taking it into the cold. Otherwise, the condensation causes icing on exposed metal parts and optics. The firing pin should be disassembled and cleaned thoroughly with a decreasing agent. It should then be lubricated with CLP. Rifle grease hardens and causes the firing pin to fall sluggishly.

g. **Salt Water Exposure.** Saltwater and saltwater atmosphere have extreme and rapid corrosive effects on the metal parts of the SWS. During periods of exposure, the SWS must be checked and cleaned as often as possible, even if it means only lubricating the SWS. The SWS should always be well lubricated, including the bore, except when actually firing. Before firing, always run a dry patch through the bore, if possible.

h. **Jungle Operations** (High Humidity). In hot and humid temperatures, keep the SWS lubricated and cased when not in use. Protect the SWS from rain and moisture whenever possible. Keep ammunition clean and dry. Clean the SWS, the bore, and the chamber daily. Keep the caps on the telescope when not in use. If moisture or fungus develops on the inside of the telescope, replace it. Clean and dry the stock daily. Dry the carrying case and SWS in the sun whenever possible.

i. **Desert Operations.** Keep the SWS dry and free of CLP and grease except on the rear of the bolt lugs. Keep the SWS free of sand by using the carrying sleeve or carrying case when not in use. Protect the SWS by using a wrap. Slide the wrap between the stock and barrel, then cross over on top of the scope. Next, cross under the SWS (over the magazine) and secure it. The SWS can still be placed into immediate operation but all critical parts are covered. The sealed hard case is preferred in the desert if the situation permits. Keep the telescope protected from the direct rays of the sun. Keep ammunition clean and protected from the direct rays of the sun. Use a toothbrush to remove sand from the bolt and receiver. Clean the bore and chamber daily. Protect the muzzle and receiver from blowing sand by covering with a clean cloth. To protect the free-floating barrel of the SWS, take an 8- or 9-inch strip of cloth and tie a knot in each end. Before going on a mission, slide the cloth between the barrel and stock all the way to the receiver and leave it there. When in position, slide the cloth out, taking all restrictive debris and sand with it.

2-4. DISASSEMBLY

Occasionally, the weapon requires disassembly however, this should be done only when absolutely necessary, not for daily maintenance. An example of this would be to remove an obstruction that is stuck between the forestock and the barrel. When disassembly is required, the recommended procedure is as follows:

a. Point the rifle in a safe direction.

b. Put the safety in the "S" position.

c. Remove the bolt assembly.

d. Use the 1/2-inch combination wrench, loosen the front and rear mounting ring nuts (Figure 2-14) on the scope, and remove the scope.

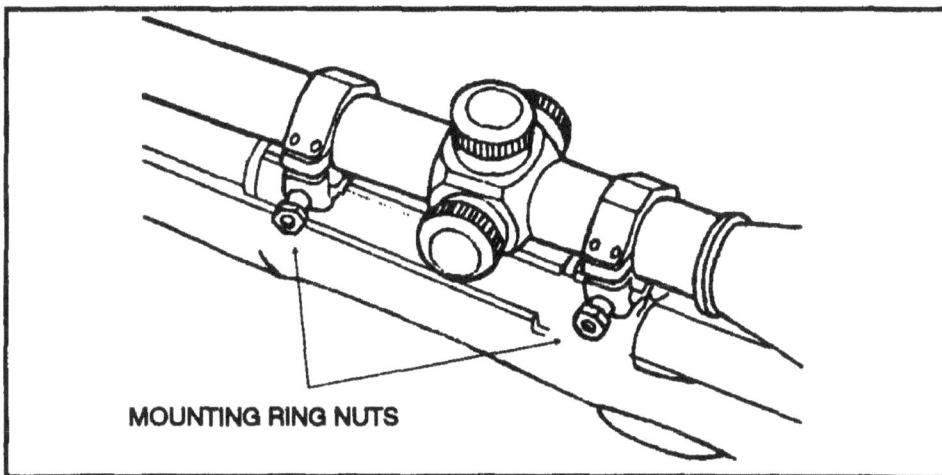

Figure 2-14. Mounting ring nuts.

e. Loosen the front and rear trigger guard screws (Figure 2-15).

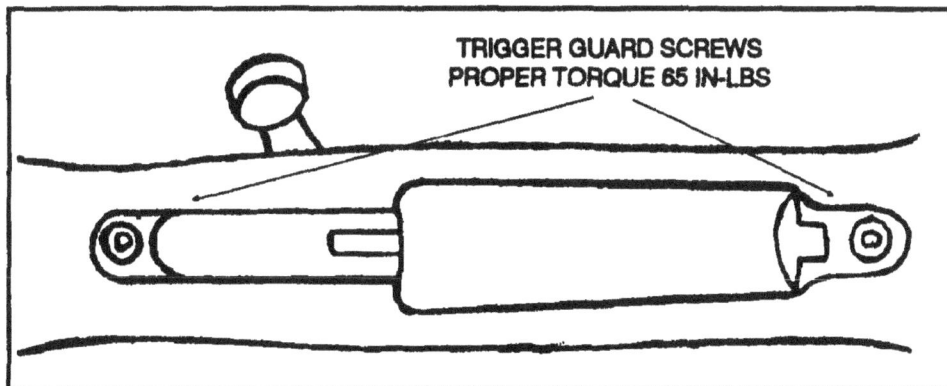

Figure 2-15. Trigger guard screws.

2-15

f. Lift the stock assembly from the barrel assembly (Figure 2-16).
g. Reassemble in reverse order.

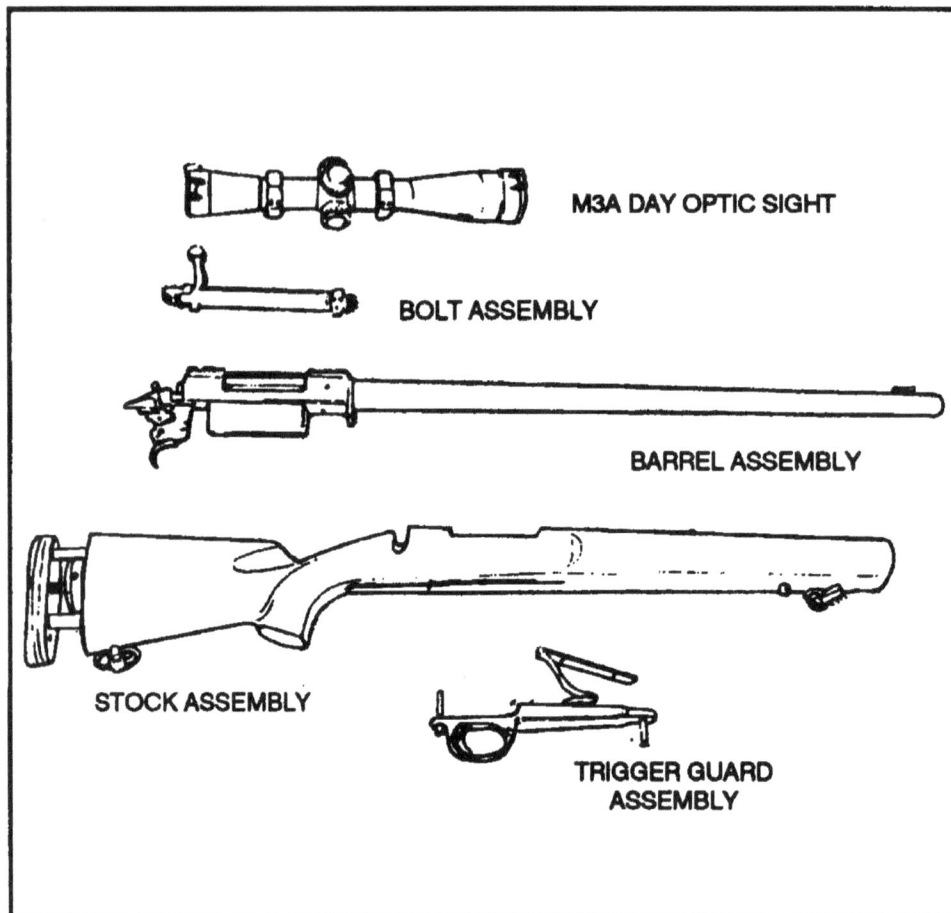

Figure 2-16. Disassembled weapon.

WARNING
ALWAYS KEEP FINGERS AWAY FROM THE TRIGGER
UNTIL READY TO FIRE, MAKE SURE THE RIFLE IS NOT
LOADED BY INSPECTING THE MAGAZINE AND CHAMBER,
USE AUTHORIZED AMMUNITION AND CHECK THE
CONDITION BEFORE LOADING THE RIFLE.

2-5. LOADING AND UNLOADING

Before loading, the sniper should ensure that the M24 SWS is on SAFE, and the bolt is in a forward position. Before unloading, he should ensure the M24 SWS is on SAFE, and the bolt is toward the rear.

a. **Loading.** The M24 has an internal, five-round capacity magazine. To load the rifle—

(1) Point the weapon in a safe direction.

(2) Ensure the safety is in the "S" position.

(3) Raise the bolt handle. Then pull the bolt handle all the way back.

(4) Push five cartridges of 7.62-mm special ball ammunition one at a time through the ejection port into the magazine. Ensure the bullet end of the cartridges is aligned toward the chamber.

(5) To ensure proper functioning, cartridges should be set fully rearward in the magazine.

(6) Use a finger to push the cartridges into the magazine and all the way down. Slowly slide the bolt forward so that the bolt slides over the top of the cartridges in the magazine.

(7) Push the bolt handle down. The magazine is now loaded.

(8) To chamber a cartridge, raise the bolt and pull it back until it stops.

(9) Push the bolt forward. The bolt removes a cartridge from the magazine and pushes it into the chamber.

(10) Push the bolt handle down.

(11) To fire, place the safety in the "F" position and squeeze the trigger.

WARNING
ENSURE THE CHAMBER AND MAGAZINE ARE CLEAR OF CARTRIDGES.

---

b. **Unloading.** To unload the M24 SWS—

(1) Point the muzzle in a safe direction.

(2) Ensure the safety is in the "S" position.

(3) Raise the bolt handle.

(4) Put one hand over the top ejection port. Slowly pull the bolt handle back with the other hand to remove the cartridge from the chamber.

(5) Remove the cartridge from the rifle.

(6) Put a hand under the floor plate.

(7) Push the floor plate latch to release the floor plate (Figure 2-17, page 2-18). The magazine spring and follower will be released from the magazine.

(8) Remove the released cartridges.

(9) Push in the magazine follower, then close the floor plate.

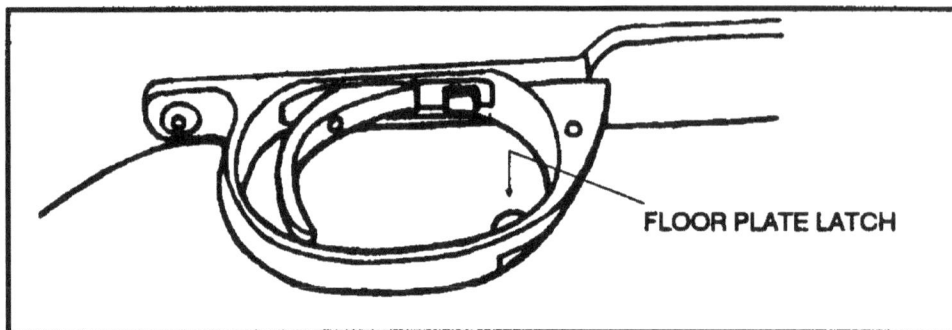

Figure 2-17. Floor plate latch.

2-6. STORAGE
The M24 SWS should be stored as follows:

a. Hang the weapon in an upside down position by the rear sling swivel.

b. Close the bolt and squeeze the trigger.

c. Open the lens caps to prevent gathering of moisture.

d. Place all other items in the system case.

e. Protect the weapon at all times during tactical movement. (See Chapter 4.)

Section II
AMMUNITION

The sniper uses the 7.62-mm special ball (M118) ammunition with the sniper weapon system. The sniper must rezero the weapon each time he fires a different type or lot of ammunition. This information should be maintained in the sniper data book.

2-7. TYPES AND CHARACTERISTICS
The types and characteristics of sniper ammunition are described in this paragraph.

a. **M118 Special Ball Bullet. The** 7.62-mm special ball (M118) bullet consists of a gilding metal jacket and a lead antimony slug. It is a boat-tailed bullet (rear of bullet is tapered) and weighs 173 grains. The tip of the bullet is not colored. The base of the cartridge is stamped with the year of manufacture and a circle that has vertical and horizontal lines, sectioning it into quarters. Its spread (accuracy standard) for a 10-shot group is no more than 12 inches at 550 meters (fired from an accuracy barrel in a test cradle).

b. **M82 Blank Ammunition.** The 7.62-mm M82 blank ammunition is used during sniper field training. It provides the muzzle blast and flash that can be detected by trainers during the exercises that evaluate the sniper's ability to conceal himself while firing his weapon.

NOTE: Regular 7.62-mm ball ammunition should be used only in an emergency situation. No damage will occur to the barrel when firing regular 7.62-mm ball ammunition. The M3A scope's bullet drop compensator is designed for M118 special ball, and there will be a significant change in zero. Therefore the rifle will not be as accurate when firing regular 7.62-mm ball ammunition. The 7.62-mm ball ammunition should be test fired and the ballistic data recorded in the data book.

2-8. ROUND-COUNT BOOK

The sniper maintains a log of the number of cartridge fired through the M24 SWS. It is imperative to accurately maintain the round-count book as the barrel should be replaced after 5,000 rounds of firing. The round-count book is issued and maintained in the arms room.

2-9. M24 MALFUNCTIONS AND CORRECTIONS

Table 2-1 does not reflect all malfunctions that can occur, or all causes and corrective actions. If a malfunction is not correctable, the complete weapon system must be turned in to the proper maintenance/ supply channel for return to the contractor (see shipment, TM 9-1005-306-10).

MALFUNCTION	CAUSE	CORRECTION
Fail to fire	Safety in "S" position	1. Move safety to "F" position
	Defective ammunition	2. Eject cartridge
	Firing pin damaged	3. Change firing pin assembly
	Firing pin binds	4. Change firing pin assembly
	Firing pin protrudes	5. Change firing pin assembly
	Firing control out of adjustment	6. Turn complete system in to the maintenance/supply channel for return to contractor
	Trigger out of adjustment	7. Turn in as above
	Trigger does not retract	8. Turn in as above
	Trigger binds on trigger guard	9. Turn in as above
	Firing pin does not remain in the cocked position with bolt closed	10. Turn in as above
Bolt binds	Action screw protudes into bolt track	11. Turn in as above
	Scope base protrudes into bolt track	12. Turn in as above

Table 2-1. M24 malfunctions and corrections.

MALFUNCTION	CAUSE	CORRECTION
Fail to feed	Bolt override of cartridge	13. Seat cartridge fully rearward in magazine
	Cartridges stems chamber	14. Pull bolt fully rearward; remove stemmed cartridge from ejection port area; reposition cartridge fully in magazine
	Magazine in backward	15. Remove magazine spring, and reinstall with long leg follower
	Weak or broken magazine spring	16. Replace spring
Fail to eject	Broken ejector	17. Turn the complete weapon system in to the maintenance/supply channel for return to contractor
	Fouled ejector plunger	18. Inspect and clean bolt face; if malfunction continues, turn in as above
Fail to extract	Broken extractor	19. Turn in as above

Table 2-1. M24 malfunctions and corrections (continued).

Section III
SNIPER SIGHTING DEVICES

The sniper has two sighting devices: the M3A scope and iron sights. The M3A scope allows the sniper to see the cross hairs and the image of the target with identical sharpness. It can be easily removed and replaced with less than 1/2 minute of angle change in zero. However, the M3A scope should be left on the rifle. Iron sights are used only as a backup sighting system and can be quickly installed.

2-10. M3A SCOPE

The M3A scope is an optical instrument that the sniper uses to improve his ability to see his target clearly in most situations. Usually, the M3A scope presents the target at an increased size (as governed by scope magnification), relative to the same target at the same distance without a scope. The M3A scope helps the sniper to identify recognize the target. His increased sighting ability also helps him to successfully engage the target.

> **NOTE: The adjustment dials are under the adjustment dust cover.**

a. **M3A Scope Adjustments.** The sniper must use the following adjustment procedures on the M3A scope:

(1) *Focus adjustment dial.* The focus adjustment dial (Figure 2-18) is on the left side of the scope barrel. This dial has limiting stops with the two extreme positions shown by the infinity mark and the largest dot. The focus adjustment dial keeps the target in focus. If the target is close, the dial is set at a position near the largest dot.

> **NOTE: Each minute of angle is an angular unit of measure.**

(2) *Elevation adjustment dial.* The elevation adjustment dial (Figure 2-18) is on top of the scope barrel. This dial has calibrated index markings from 1 to 10. These markings represent the elevation setting adjustments needed at varying distances: 1 = 100 meters, 3 = 300 meters, 7 = 700 meters, and so on. Each click of the elevation dial equals 1 minute of angle.

(3) *Windage adjustment dial.* The windage adjustment dial (Figure 2-18) is on the right side of the scope barrel. This dial is used to make lateral adjustments to the scope. Turning the dial in the indicated direction moves the point of impact in that direction. Each click on the windage dial equals .5 minute of angle.

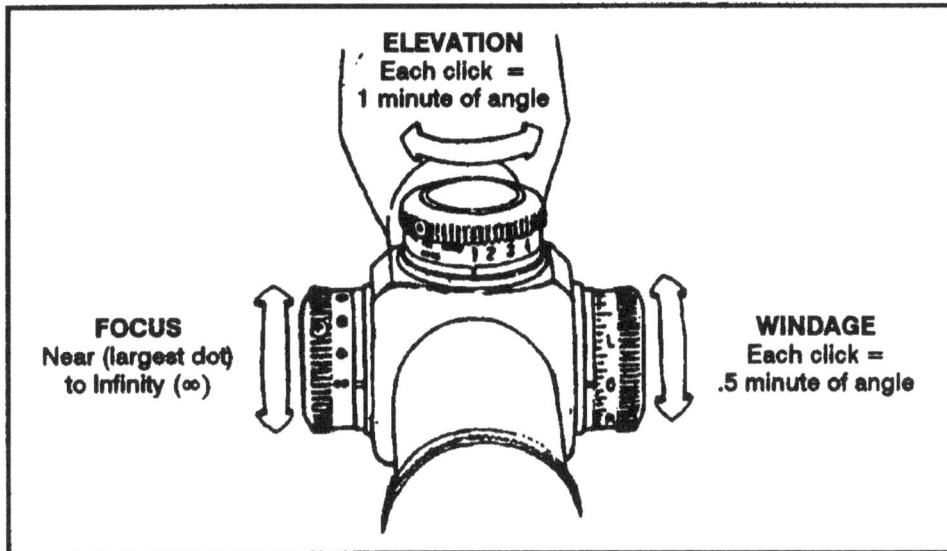

Figure 2-18. Focus, elevation, and windage adjustment dials.

(4) *Eyepiece adjustment.* The eyepiece (Figure 2-19) is adjusted by turning it in or out of the barrel until the reticle appears crisp and clear. Focusing the eyepiece should be done after mounting the scope. The sniper grasps the eyepiece and backs it away from the lock ring. He does not attempt to loosen the lock ring first; it loosens automatically when he backs away from the eyepiece (no tools needed). The eyepiece is turned several turns to move it at least 1/8 inch. It takes this much change to achieve any measurable effect on the focus. The sniper looks through the scope at the sky or a blank wall and checks to see if the reticle appears sharp and crisp. He locks the lock ring after achieving reticle clarity.

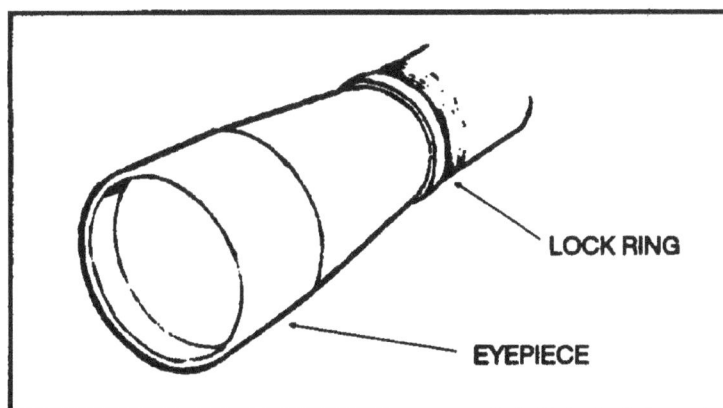

Figure 2-19. Eyepiece adjustment.

WARNINGS

1. SECURELY FASTEN THE MOUNTING BASE TO THE RIFLE. LOOSE MOUNTING MAY CAUSE THE M3A SCOPE AND BASE MOUNT ASSEMBLY TO COME OFF THE RIFLE WHEN FIRING, POSSIBLY INJURING THE FIRER.

2. DURING RECOIL PREVENT THE M3A SCOPE FROM STRIKING THE FACE BY MAINTAINING AN AVERAGE DISTANCE OF 2 TO 3 INCHES BETWEEN THE EYE AND THE SCOPE.

b. **M3A Scope Mount.** The M3A scope mount has a baseplate with four screws; a pair of scope rings with eight ring screws, each with an upper and lower ring half with eight ring screws and two ring mounting bolts with nuts (Figure 2-20). The baseplate is mounted to the rifle by screwing the four baseplate screws through the plate and into the top of the receiver. The screws must not protrude into the receiver and interrupt the functioning of the bolt. After the baseplate is mounted, the scope rings are mounted.

NOTE: The M3A scope has two sets of mounting slots. The sniper selects the set of slots that provides proper eye relief (the distance that the eye is positioned behind the telescopic sight). The average distance is 2 to 3 inches. The sniper adjusts eye relief to obtain a full field of view.

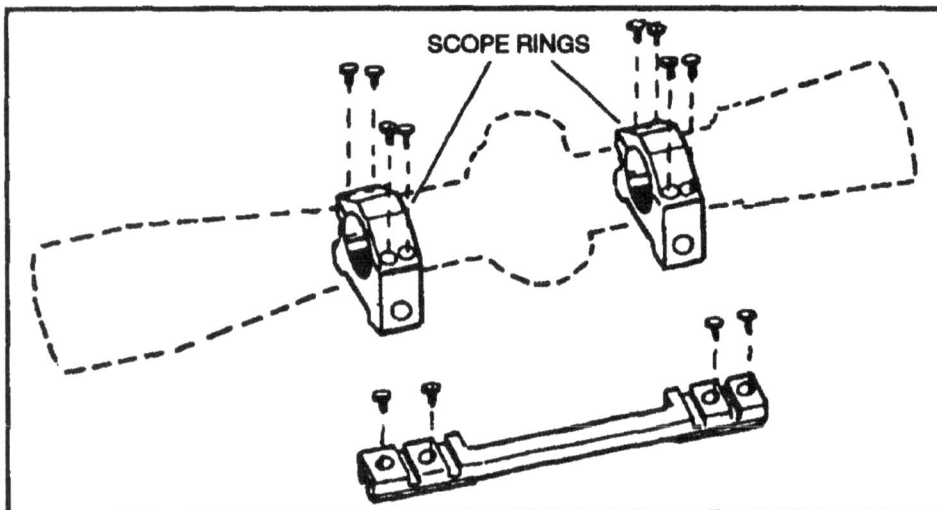

Figure 2-20. Scope mount.

(1) Before mounting the M3A scope, lubricate the threads of each mounting ring nut.

(2) Ensure smooth movement of each mounting ring nut and mount claw.

(3) Inspect for burrs and foreign matter between each mounting ring nut and mount claw. Remove burrs or foreign matter before mounting.

(4) Mount the sight and rings to the base.

NOTE: Once a set of slots is chosen, the same set should always be used in order for the SWS to retain zero.

(5) Ensure the mounting surface is free of dirt, oil, or grease.

(6) Set each ring bolt spline into the selected slot.

(7) Slide the rear mount claw against the base and finger-tighten the mounting ring nut.

(8) If the scope needs to be adjusted loosen the mounting ring nuts and align the ring bolts with the other set of slots on the base Repeat this process.

(9) Slide the front mount claw against the base, and finger-tighten the mounting ring nut.

(10) Use the T-handle torque wrench, which is preset to 65inch-pounds, to tighten the rear mounting ring nut.

c. **Care and Maintenance of the M3A Scope.** Dirt, rough handling, or abuse of optical equipment will result in inaccuracy and malfunction. When not in use, the rifle and scope should be cased, and the lens should be capped.

(1) *Lens.* The lens are coated with a special magnesium fluoride reflection-reducing material. This coat is thin and great care is required to prevent damage to it.

(a) To remove dust, lint, or other foreign matter from the lens, lightly brush the lens with a clean camel's-hair brush.

(b) To remove oil or grease from the optical surfaces, apply a drop of lens cleaning fluid or robbing alcohol on a lens tissue. Carefully wipe off the surface of the lens in circular motions (from the center to the outside edge). Dry off the lens with a clean lens tissue. In the field, if the proper supplies are not available, breathe heavily on the glass and wipe with a soft, clean cloth.

(2) *Scope.* The scope is a delicate instrument and must be handled with care. The following precautions will prevent damage

(a) Check and tighten all mounting screws periodically and always before an operation. Be careful not to change the coarse windage adjustment.

(b) Keep the lens free from oil and grease and never touch them with the fingers. Body grease and perspiration can injure them. Keep the cap on the lens.

(c) Do not force the elevation and windage screws or knobs.

(d) Do not allow the scope to remain in direct sunlight, and avoid letting the sun's rays shine through the lens. The lens magnify and concentrate sunlight into a pinpoint of intense heat, which is focused on the mil-scale reticle. This may melt the mil dots and damage the scope internally. Keep the lens covered and the entire scope covered when not in use.

(e) Avoid dropping the scope or striking it with another object. This could permanently damage the telescope as well as change the zero.

(f) To avoid damage to the scope or any other piece of sniper equipment, snipers or armorers should be the only personnel handling the equipment. Anyone who does not know how to use this equipment could cause damage.

(3) *Climate conditions.* Climate conditions play an important part in taking care of optical equipment.

(a) *Cold climates.* In extreme cold, care must be taken to avoid condensation and congealing of oil on the glass of the optical equipment. If the temperature is not excessive, condensation can be removed by placing the instrument in a warm place. Concentrated heat must not be applied because it causes expansion and damage can occur. Moisture may also be blotted from the optics with lens tissue or a soft, dry cloth. In cold temperatures, oil thickens and causes sluggish operation or failure. Focusing parts are sensitive to freezing oils. Breathing forms frost, so the optical surfaces must be cleaned with lens tissue, preferably dampened lightly with alcohol. DO NOT apply alcohol on the glass of the optics.

(b) *Jungle operations (high humidity).* In hot and humid temperatures, keep the caps on the scope when not in use. If moisture or fungus develops on the inside of the telescope, replace it.

(c) *Desert operations.* Keep the scope protected from the direct rays of the sun.

(d) *Hot climate and salt water exposure.* The scope is vulnerable to hot, humid climates and salt water atmosphere. It MUST NOT be exposed to direct sunlight. In humid and salt air conditions, the scope must be inspected, cleaned, and lightly oiled to avoid rust and corrosion. Perspiration can also cause the equipment to rust; therefore, the instruments must be thoroughly dried and lightly oiled.

d. **M3A Scope Operation.** When using the M3A scope, the sniper looks at the target and determines the distance to it by using the mil dots on the reticle. The mil-dot reticle (Figure 2-21) is a duplex-style reticle that has thick outer sections and thin inner sections. Superimposed on the thin center section of the reticle is a series of dots. There are 4 dots on each side of the center and 4 dots above and below the center. These 4 dots are spaced 1 mil apart, and 1 mil from both the center and the start of the thick section of the reticle. This spacing allows the sniper to make close estimates of target range, assuming there is an object of known size (estimate) in the field of view. For example, a human target appears to be 6 feet tall, which equals 1.83 meters tall, and at 500 meters, 3.65 dots high (nominally, about 3.5 dots high). Another example is a l-meter target at a 1,000-meter range. This target is the height between 2 dots, or the width between 2 dots. If the sniper is given a good estimate of the object's size, then he may accurately determine target range using the mil-dot system.

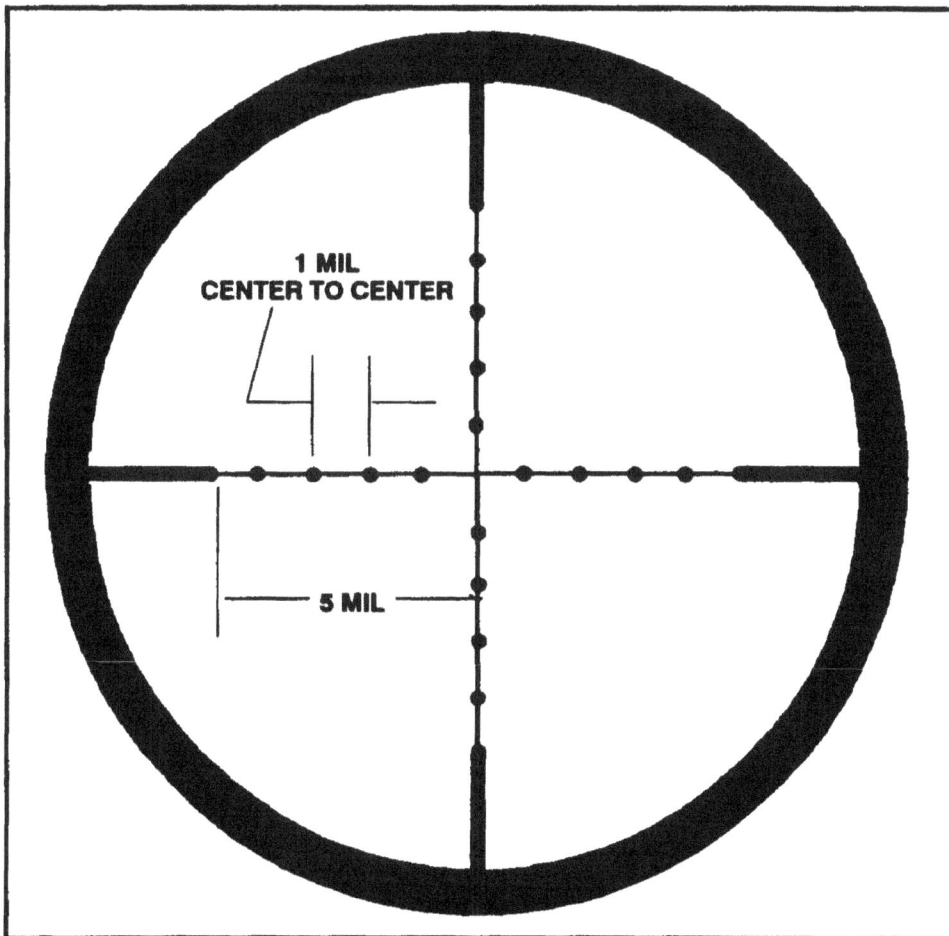

Figure 2-21. Mil-dot reticle.

e. **Zeroing.** Zeroing the M3A scope should be done on a known-distance range (preferably 900 meters long) with bull's-eye-type targets (200-yard targets, NSN SR1-6920-00-900-8204). When zeroing the scope, the sniper—

(1) Assumes a good prone-supported position 100 meters from the target.

(2) Ensures the "l" on the elevation dial is lined up with the elevation index line, and the "0" on the windage dial is lined up with the windage index line.

(3) Fires three rounds at the center of the target, keeping the same aiming point each time and triangulate.

(4) After the strike of the rounds has been noted, turns the elevation and windage dials to make the needed adjustments to the scope.

- Each click on the elevation dial equals one minute of angle.
- One minute of angle at 100 meters equals 1.145 inches or about 1 inch.
- Each click on the windage dial equals .5 minute of angle.
- .5 minute of angle at 100 meters equals about .5 inch.

(5) Repeats steps 3 and 4 until a three-round shot group is centered on the target.

(6) Once the shot group is centered, loosens the hex head screws on the elevation and windage dials. He turns the elevation dial to the index line marked "l" (if needed). He turns the windage dial to the index line marked "0" (if needed) and tighten the hex head screws.

(7) After zeroing at 100 meters and calibrating the dial, confirms this zero by firing and recording sight settings (see Chapter 3) at 100-meter increments through 900 meters.

f. **Field-Expedient Confirmation/Zeroing.** The sniper may need to confirm zero in a field environment. Examples are shortly after receiving a mission, a weapon was dropped, or excessive climatic changes as may be experienced by deploying to another part of the world. Two techniques of achieving a crude zero are the 25-yard/ 900-inch method and the observation of impact method.

(1) *25-yard/900-inch method.* Dial the scope to 300 meters for elevation and to "0" for windage. Aim and fire at a target that is at a 25-yard distance. Adjust the scope until rounds are impacting 5/ 8 of an inch above the point of aim. To confirm, set the elevation to 500 meters. The rounds should impact 2 1/ 4 inches above the point of aim.

(2) *Observation of impact method.* When a known distance range is unavailable, locate a target so that the observer can see the impact of

rounds clearly. Determine the exact range to the target, dial in the appropriate range, and fire. Watch the impact of the rounds; the observer gives the sight adjustments until a point of aim or point of impact is achieved.

2-11. IRON SIGHTS

Depending on the situation, a sniper may be required to deliver an effective shot at ranges up to 900 meters or more. This requires the sniper to zero his rifle with the iron sights and the M3A scope at most ranges that he can be expected to fire.

a. **Mounting.** To mount iron sights, the sniper must remove the M3A scope first.

(1) Attach the front sight to the barrel, align the front sight and the front sight base, and slide the sight over the base and tighten the screw (Figure 2-22).

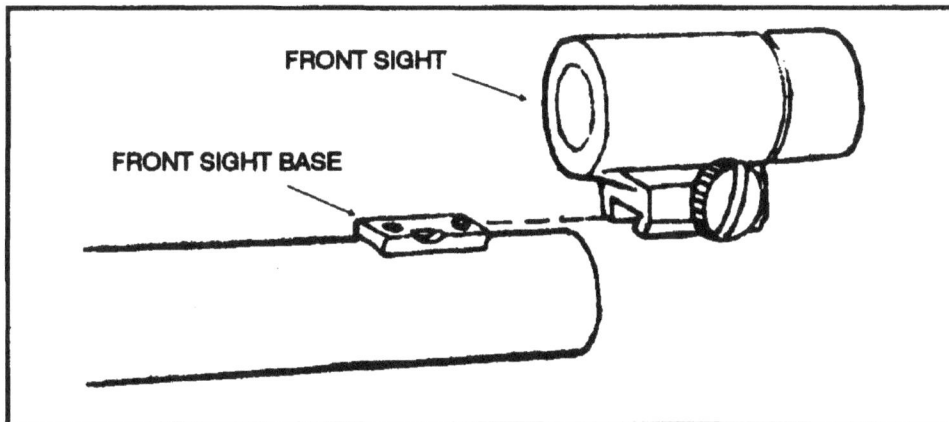

Figure 2-22. Front sight attachment.

(2) The aperture insert may be either skeleton or translucent plastic (Figure 2-23, page 2-30). The skeleton aperture is the most widely used. The translucent plastic aperture is preferred by some shooters and is available in clear plastic. Both apertures are available in various sizes. A common error is selecting an aperture that is too small. Select an aperture that appears to be at least twice the diameter of the bull's-eye. An aperture selected under one light condition may, under a different light, form a halo around the bull or make the bull appear indistinct or oblong. The aperture selected should reveal a wide line of white around the bull and allow the bull to standout in clear definition against this background.

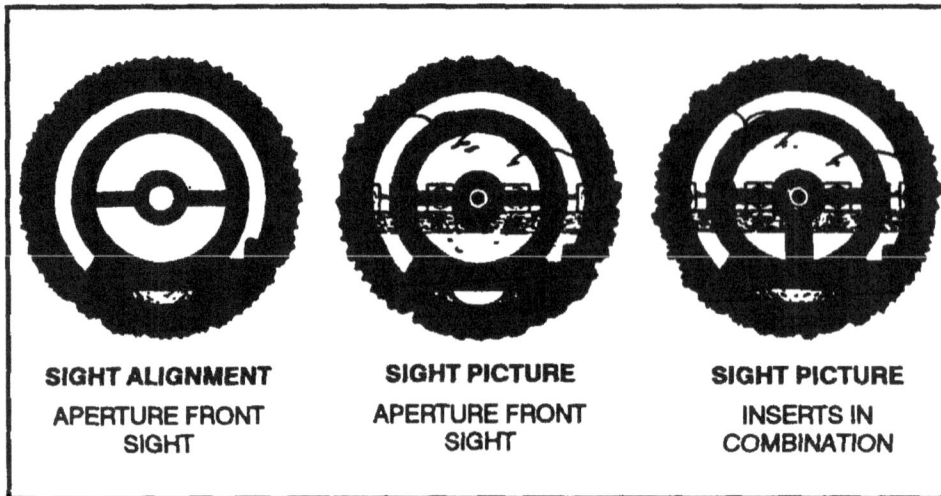

Figure 2-23. Aperture insert.

(3) Remove one of the three sets of screws from the rear sight base located on the left rear of the receiver. Align the rear sight with the rear sight base taking care to use the hole that provides the operator the desired eye relief. Then tighten the screw to secure the rear sight to the base.

NOTE: Operator-desired eye relief determines the set screw that must be removed.

b. **Adjustment Scales.** Adjustment scales are of the vernier type. Each graduation on the scale inscribed on the sight base equals 3 minutes of angle. (See the minutes of angle chart in Chapter 3.) Each graduation of the adjustable scale plates equals 1 minute of angle. To use the vernier-type adjustment scales—

(1) Note the point at which graduations on both the top and the bottom scales are aligned.

(2) Count the numbers of full 3 minutes of angle graduations from "0" on the fixed scale to "0" on the adjustable scale. Add this figure to the number of 1 minute of angle graduations from "0" on the adjustable scale to the point where the two graduations are aligned.

c. Zeroing. Zeroing iron sights should be done on the same type of range and targets as in paragraph 2-10a. To set a mechanical zero on the iron sights for windage, the sniper turns the windage dial all the way to the left or right, then he counts the number of clicks it takes to get from one side to the other. He divides this number by 2—for example, 120 divided by 2 equals 60. The sniper turns the windage dial 60 clicks back to the center. If the two zeros on the windage indicator plate do not align, he loosens the screw on the windage indicator plate and aligns the two zeros. The sniper uses the same procedure to set a mechanical zero for elevation. Once a mechanical zero has been set, he assumes a good prone-supported position, 100 meters from the target. He fires three rounds at the center of the target, observing the same aiming point each time. After noting the strike of the rounds, the sniper turns the *elevation* and *windage* dials to make needed adjustments to the iron sights as follows (Figure 2-24):

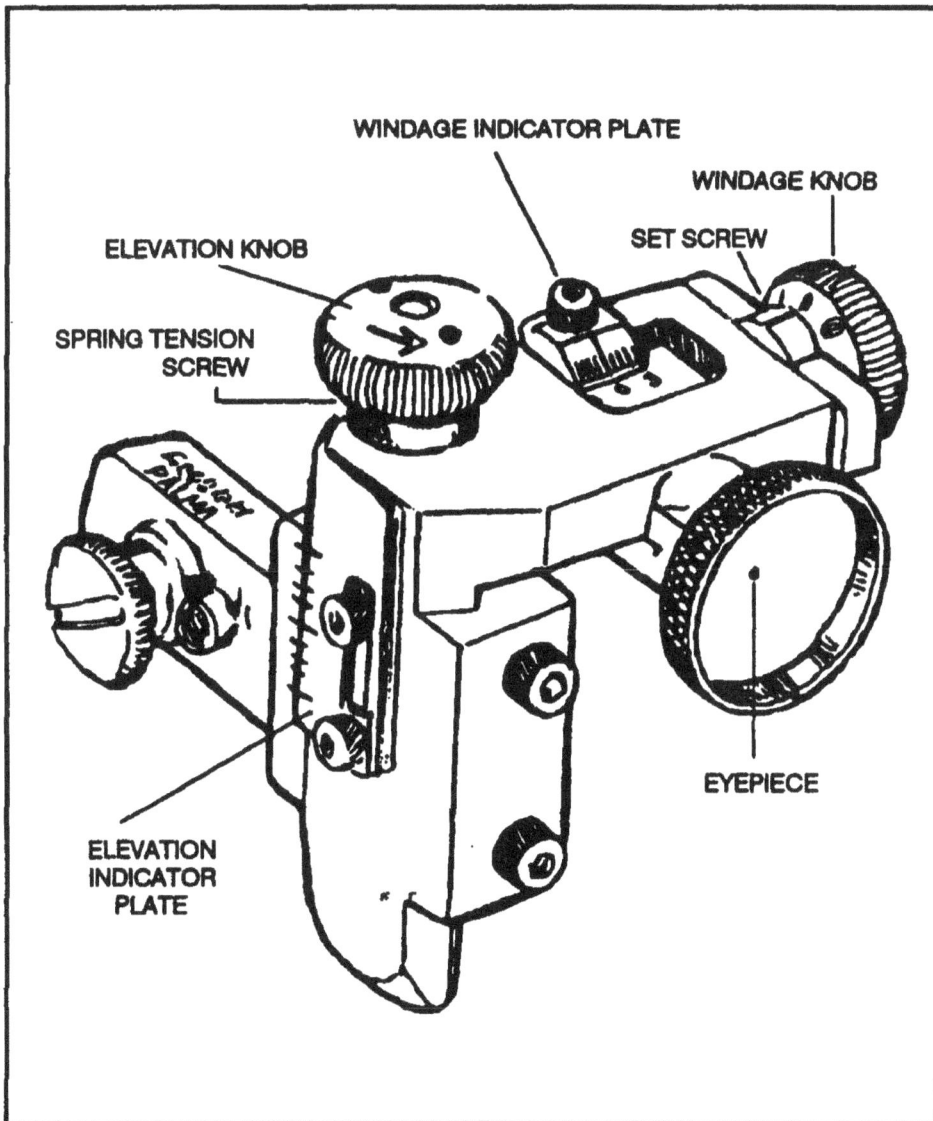

Figure 2-24. Zeroing adjustment dials.

(1) Each click of adjustment is 1/4 minute of angle (one minute of angle equals about 1 inch at 100 yards, 6 inches at 600 yards, and so forth). There are twelve 1/4 minutes of angle, equaling 3 minutes of angle adjustments in each dial revolution. The total elevation adjustment is 60 minutes of angle (600 inches at 1,000 yards) total windage adjustment is 36 minutes of angle (360 inches at 1,000 yards).

(2) Turn the elevation dial in the direction marked *UP* to raise the point of impact: turn the elevation dial in the opposite direction to lower the point of impact. Turn the windage dial in the direction marked *R* to move the point of impact to the right; then turn the windage dial in the opposite direction to move the point of impact to the left.

(3) Continue firing and adjusting shot groups until the point of aim or point of impact is achieved.

After zeroing the rifle sight to the preferred range, the sniper loosens the elevation and windage indicator plate screws with the socket head screw key provided. Now, he loosens the spring tension screw, aligns the "0" on the plate with the "0" on the sight body, and retightens the plate screws. Then the sniper loosens the spring tension screws and set screws in each dial, and aligns the "0" of the dial with the reference line on the sight. He presses the dial against the sight, tightens the set screws, and equally tightens the spring tension screws until a definite "click" can be felt when the dial is turned. This click can be sharpened or softened to preference by equally loosening or tightening the spring screws on each dial. The sniper makes windage and elevation corrections, and returns quickly to "zero" standard.

Section IV
OTHER EQUIPMENT

The sniper must use special equipment to reduce the possibility of detection. The types and characteristics are discussed in this section.

2-12. M16A1/A2 RIFLE WITH M203 GRENADE LAUNCHER

The observer carries the M16A1/A2 rifle with the M203 grenade launcher. The sniper, carrying the M24 SWS, lacks the firepower required to break contact with enemy forces-that is, ambush or chance contact. The rapid-fire ability of the M16A1/A2 rifle, combined with the destructive abilities of the M203 40-mm grenade launcher (Figure 2-25), gives the sniper team a lightweight, easily operated way to deliver the firepower required to break contact. (See FM 23-9 and FM 23-31, respectively, for the technical characteristics of these weapons.)

Figure 2-25. The M203 40-mm grenade launcher attached to M16A1 rifle.

2-13. IMAGE INTENSIFICATION AND INFRARED DEVICES

The sniper team employs night and limited visibility devices to conduct continuous operations.

a. **Night Vision Sight, AN/PVS-4.** The AN/ PVS-4 is a portable, battery-operated, electro-optical instrument that can be hand-held for visual observation or weapon-mounted for precision fire at night (Figure 2-26). The observer can detect and resolve distant targets through the unique capability of the sight to amplify reflected ambient light (moon, stars, or sky glow). The sight is passive thus, it is free from enemy detection by visual or electronic means. This sight, with appropriate weapons adapter bracket, can be mounted on the M16 rifle.

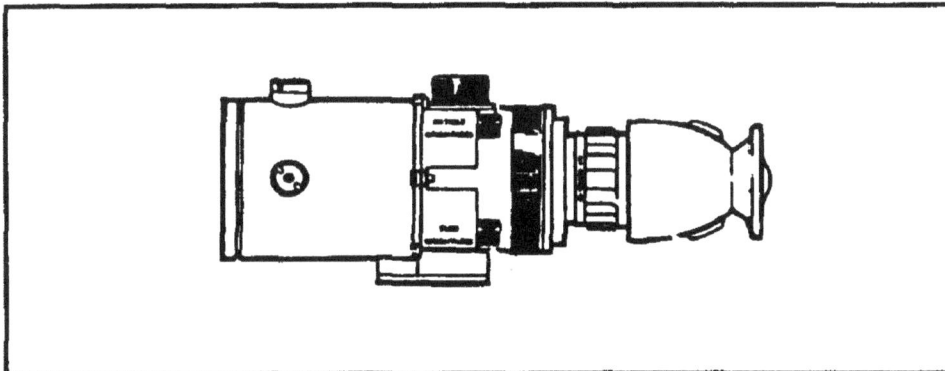

Figure 2-26. Night vision sight, AN/PVS-4.

(1) *Uses.* The M16 rifle with the mounted AN/ PVS-4 is effective in achieving a first-round hit out to and beyond 300 meters, depending on the light conditions. The AN/ PVS-4 is mounted on the M16 since the

nightsight's limited range does not make its use practical for the sniper weapon system. This avoids problems that may occur when removing and replacing the sniperscope. The nightsight provides an effective observation ability during night combat operations. The sight does not give the width, depth, or clarity of daylight vision; however, a well-trained operator can see enough to analyze the tactical situation, to detect enemy targets, and to place effective fire on them. The sniper team uses the AN/PVS-4 to accomplish the following.

(a) To enhance their night observation capability.

(b) To locate and suppress hostile fire at night.

(c) To deny enemy movement at night.

(d) To demoralize the enemy with effective first-round kills at night.

(2) *Employment factors.* Since the sight requires target illumination and does not project its own light source, it will not function in total darkness. The sight works best on a bright, moonlit night. When there is no light or the ambient light level is low (such as in heavy vegetation), the use of artificial or infrared light improves the sight's performance.

(a) Fog, smoke, dust, hail, or rain limit the range and decrease the resolution of the instrument.

(b) The sight does not allow seeing through objects in the field of view. For example, the operator will experience the same range restrictions when viewing dense wood lines as he would when using other optical sights.

(c) The observer may experience eye fatigue when viewing for prolonged periods. Viewing should be limited to 10 minutes, followed by a rest period of 10 minutes. After several periods of viewing, he can safely extend this time limit. To assist in maintaining a continuous viewing capability and to reduce eye fatigue, the observer should use one eye then the other while viewing through the sight.

(3) **Zeroing.** The operator may zero the sight during daylight or darkness; however, he may have some difficulty in zeroing just l before darkness. The light level at dusk is too low to permit the operator to resolve his zero target with the lens cap cover in place, but it is still intense enough to cause the sight to automatically turnoff unless the lens cap cover is in position over the objective lens. The sniper normally zeros the sight for the maximum practical range that he can be expected to observe and fire, depending on the level of light.

b. **Night Vision Goggles, AN/PVS-5.** The AN/PVS-5 is a lightweight, passive night vision system that gives the sniper team another means of observing an area during darkness (Figure 2-27). The sniper normally carries the goggles, because the observer has the M16 mounted with the nightsight. The goggles make it easier to see due to their design. However, the same limitations that apply to the nightsight also apply to the goggles.

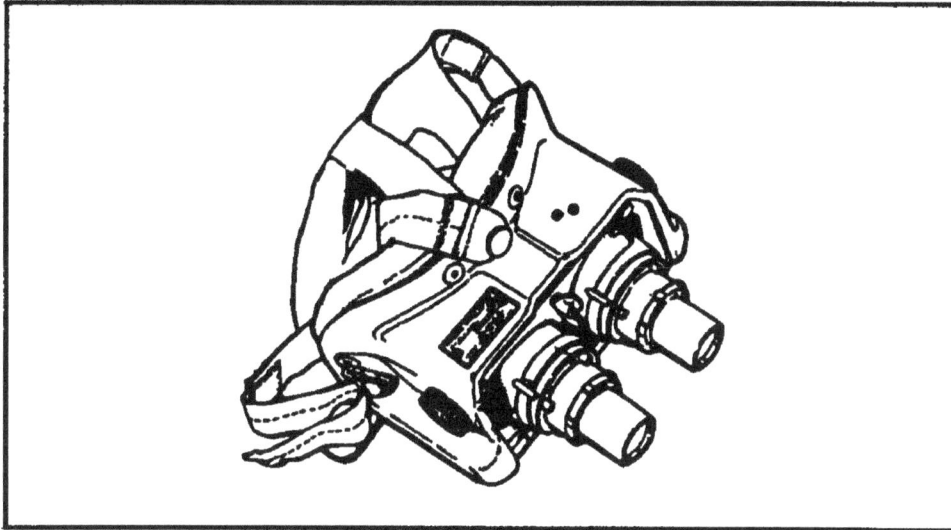

Figure 2-27. Night vision goggles, AN/PVS-5.

c. **Night Vision Goggles, AN/PVS-7 Series.** The night vision goggles, AN/ PVS-7 series (Figure 2-28, page 2-36) has a better resolution and viewing ability than the AN/ PVS-5 goggles. The AN/ PVS-7 series goggles have a head-mount assembly that allows them to be mounted in front of the face so that both hands can be free. The goggles can be used without the mount assembly for hand-held viewing. (See TM 11-5855-262-10-1.)

d. **Laser Observation Set AN/GVS-5.** Depending on the mission, snipers can use the AN/ GVS-5 to determine the range to the target. The AN/ GVS-5 (LR) (Figure 2-29, page 2-36) is an individually operated, hand-held, distance-measuring device designed for distances from 200 to 9,990 meters (with an error of plus or minus 10 meters). It measures distances by firing an infrared beam at a target and by measuring the time the reflected beam takes to return to the operator. It then displays the target distance, in meters, inside the viewer. The reticle pattern in the viewer is graduated in 10-mil increments and has display lights to indicate low battery and multiple target hits. If the beam hits more than one target, the display gives a reading of the closest target hit. The beam that is fired from the set poses a safety hazard; therefore, snipers planning to use this equipment should be thoroughly trained in its safe operation. (See TM 11-5860-201-10.)

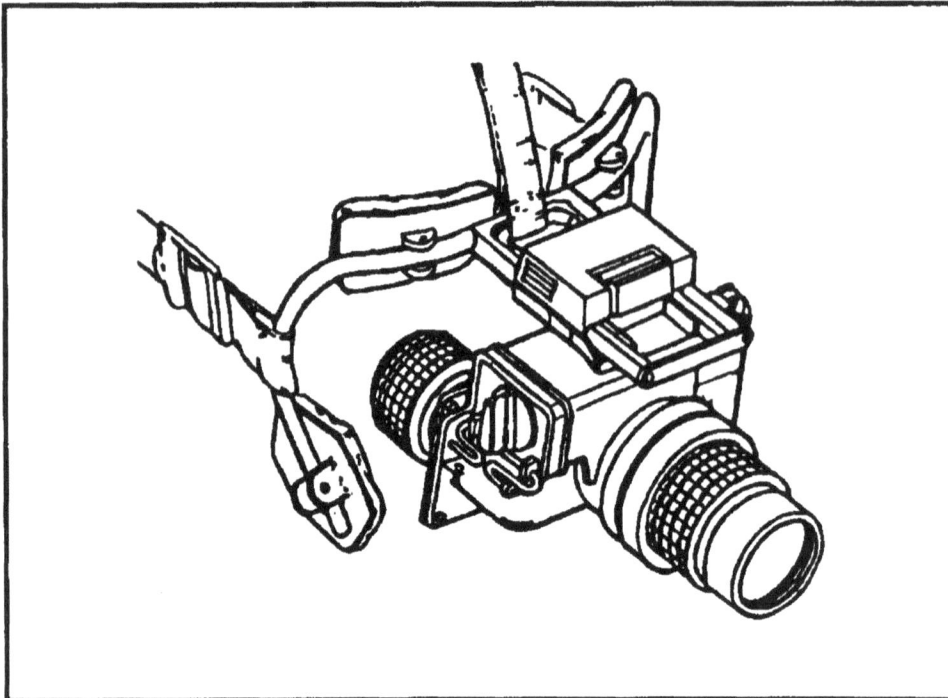

Figure 2-28. Night vision goggles, AN/PVS-7 series.

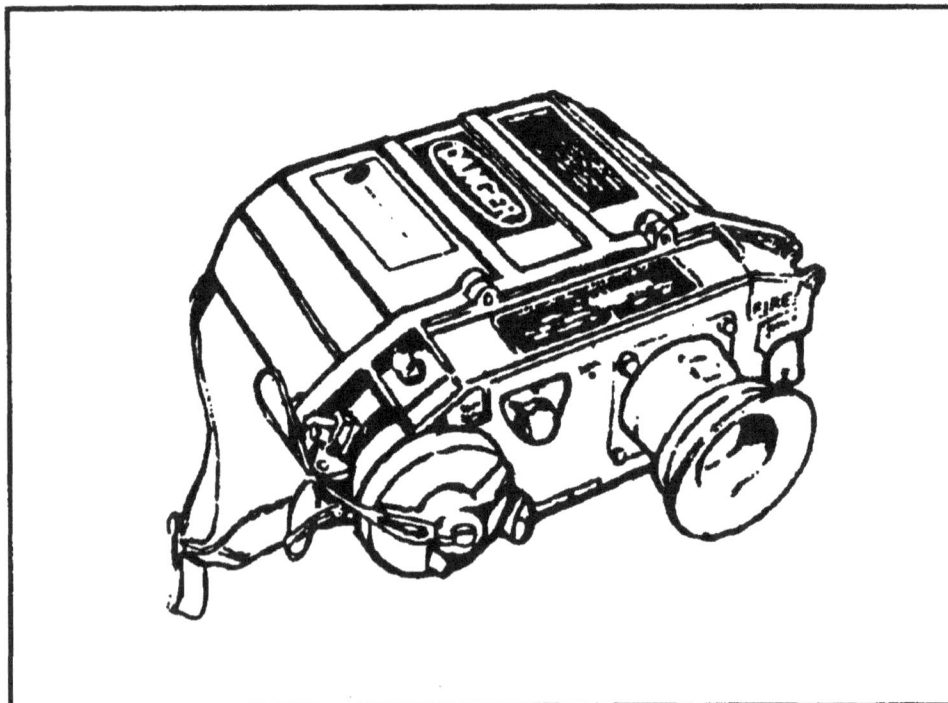

Figure 2-29. Laser observation set, AN/GVS-5.

e. **Mini-Eyesafe Laser Infrared Observation Set, AN/PVS-6.**
The AN/ PVS-6 (Figure 2-30) contains the following components:
mini-eyesafe laser range finder; batteries, BA-6516/ U, nonrechargeable,
lithium thionyl chloride;
carrying case; shipping case;
tripod; lens cleaning compound and lens cleaning
tissue; and operator's manual.
The laser range finder is the
major component of the
AN/ PVS-6. It is lightweight,
individually operated, and
hand-held or tripod mounted;
it can accurately determine
ranges from 50 to 9,995 meters
in 5-meter increments and
displays the range in the
eyepiece. It can also be
mounted with and bore-
sighted to the night observation device, AN/ TAS-6,
long-range.

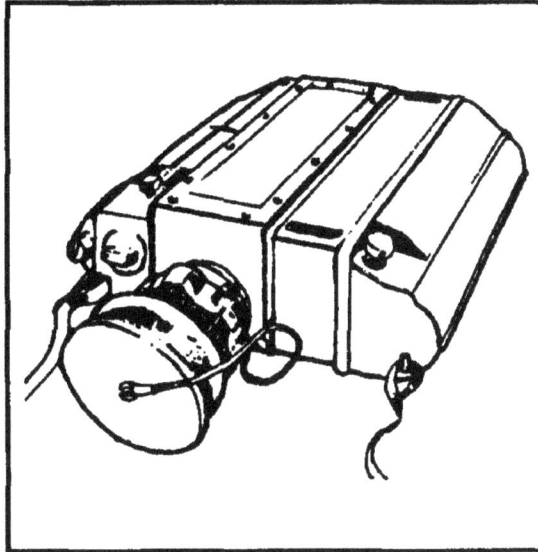

**Figure 2-30. Mini-eyesafe laser
infrared observation set,
AV/PVS-6.**

2-14. M49 OBSERVATION TELESCOPE

The M49 observation telescope is a prismatic optical instrument of
20-power magnification (Figure 2-31, page 2-38). The telescope is
focused by turning the eyepiece in or out until the image of the object
being viewed is crisp and clear to the viewer. The sniper team carries the
telescope on all missions. The observer uses the telescope to determine
wind speed and direction by reading mirage, observing the bullet trace,
and observing the bullet impact. The sniper uses this information to
make quick and accurate adjustments for wind conditions. The lens are
coated with a hard film of magnesium fluoride for maximum light
transmission. Its high magnification makes observation, target detection,
and target identification possible where conditions and range would
otherwise preclude this ability. Camouflaged targets and those in deep
shadows can be more readily distinguished. The team can observe troop
movements at greater distances and identify selective targets with ease.

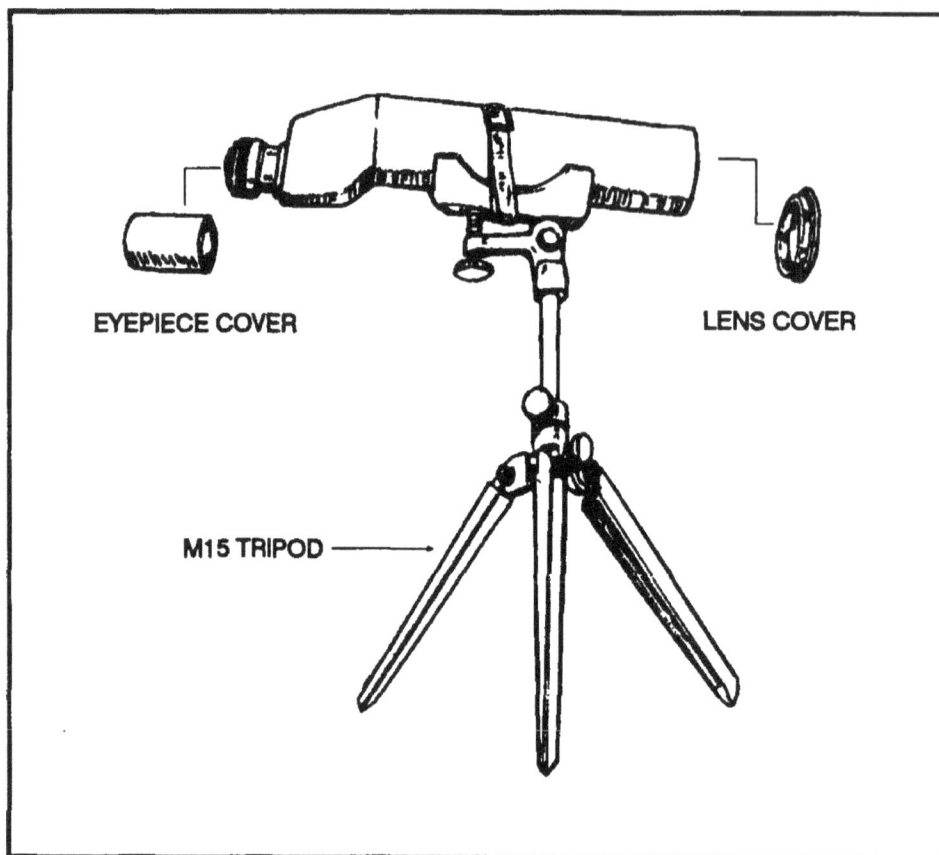

Figure 2-31. M49 observation telescope.

a. **Components.** Components of the telescope include a removable eyepiece and objective lens covers, an M15 tripod with canvas carrier, and a hard ease carrier for the telescope.

b. **Storage.** When storing the M49 observation telescope, the sniper must remove it from the hard case earner and remove the lens caps to prevent moisture from gathering on the inside of the scope. Maintenance consists of—

(1) Wiping dirt and foreign materials from the scope tube, hard case carrier, and M15 tripod with a damp rag.

(2) Cleaning the M49 lens with lens cleaning solution and lens tissue only.

(3) Brushing dirt and foreign agents from the M15 carrying case with a stiff-bristled brush; cleaning the threading of lens caps on the M49 and the tripod elevation adjustment screw on the M15 with a toothbrush, then applying a thin coat of grease and moving the lens caps and elevation adjustment screw back and forth to evenly coat threading.

2-15. M19 BINOCULARS

The M19 is the preferred optical instrument for conducting hasty scans. This binocular (Figure 2-32) has 7-power magnification with a 50-mm objective lens, and an interpupillary scale located on the hinge. The sniper should adjust the binocular until one sharp circle appears while looking through them. After adjusting the binoculars' interpupillary distance (distance between a person's pupils), the sniper should make a mental note of the reading on this scale for future reference. The eyepieces are also adjustable. The sniper can adjust one eyepiece at a time by turning the eyepiece with one hand while placing the palm of the other hand over the objective lens of the other monocular. While keeping both eyes open, he adjusts the eyepiece until he can see a crisp, clear view. After one eyepiece is adjusted, he repeats the procedure with the remaining eyepiece. The sniper should also make a mental note of the diopter scale reading on both eyepieces for future reference. One side of the binoculars has a laminated reticle pattern (Figure 2-32) that consists of a vertical and horizontal mil scale that is graduated in 10-mil increments. Using this reticle pattern aids the sniper in determining range and adjusting indirect-fires. The sniper uses the binoculars for—

- Calling for and adjusting indirect fires.
- Observing target areas.
- Observing enemy movement and positions.
- Identifying aircraft.
- Improving low-light level viewing.
- Estimating range.

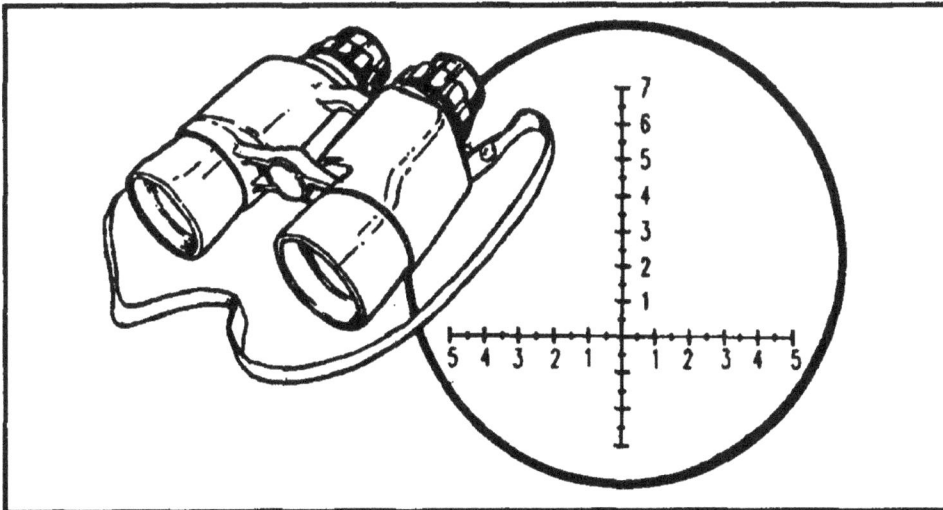

Figure 2-32. M19 binoculars and reticle.

2-16. M22 BINOCULARS

The M22 binoculars (Figure 2-33) can be used instead of the M19. These binoculars have the same features as the M19, plus fold-down eyepiece cups for personnel who wear glasses to reduce the distance between the eyes and the eyepiece. It also has protective covers for the objective and eyepiece lenses. The binoculars have laser protection filters on the inside of the objective lenses (direct sunlight can reflect off these lenses). The reticle pattern (Figure 2-33) is different than the M19 binocular reticle.

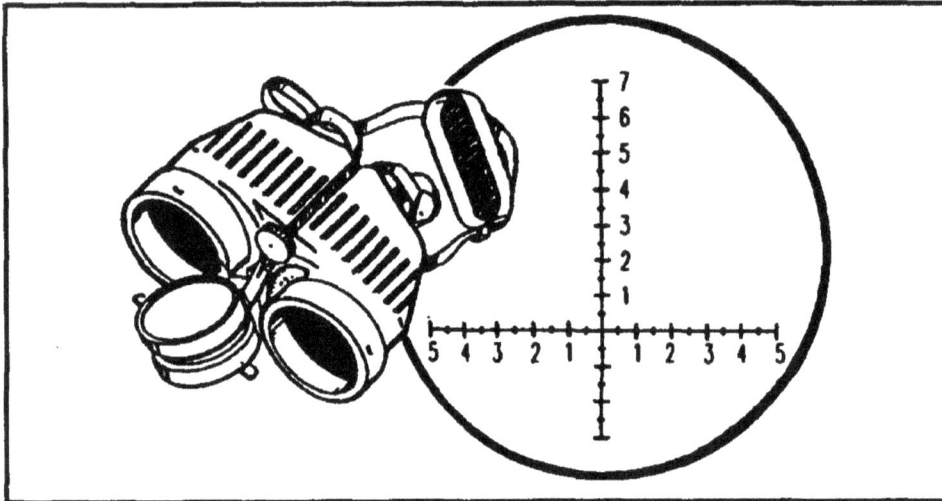

Figure 2-33. M22 binoculars and reticle.

2-17. OTHER SNIPER EQUIPMENT

Other equipment the sniper needs to complete a successful mission follows:

a. **Sidearms.** Each member of the team should have a sidearm, such as an M9, 9-mm Beretta, or a caliber .45 pistol. A sidearm gives a sniper the needed protection from a nearby threat while on the ground moving or while in the confines of a sniper position.

b. **Compass.** Each member of the sniper team must have a lensatic compass for land navigation.

c. **Maps.** The team must have military maps of their area of operations.

d. **Calculator.** The sniper team needs a pocket-size calculator to figure distances when using the mil-relation formula. Solar-powered calculators usually work well, but under low-light conditions, battery power may be preferred. If a battery-powered calculator is to be used in low-light conditions, it should have a lighted display.

e. **Rucksack.** The sniper's rucksack should contain at least a two-quart canteen, an entrenching tool, a first-aid kit, pruning shears, a sewing kit with canvas needles and nylon thread, spare netting and garnish, rations, and personal items as needed. The sniper also carries his ghillie suit (Chapter 4, paragraph 4-4) in his rucksack until the mission requires its use.

f. **Measuring Tape.** A standard 10-foot to 25-foot metal carpenter's tape allows the sniper to measure items in his operational area. This information is recorded in the sniper data book. (See Chapter 4 for range estimation.)

Section V
COMMUNICATIONS EQUIPMENT

The sniper team must have a man-portable radio that gives the team secure communications with the units involved in their mission.

2-18. AN/PRC-77 RADIO

The basic radio for the sniper team is the AN/ PRC-77 (Figure 2-34). This radio is a short-range, man-pack portable, frequency modulated receiver-transmitter that provides two-way voice communication. The set can net with all other infantry and artillery FM radio sets on common frequencies. The AN/ KY-57 should be installed with the AN/ PRC-77. This allows the sniper team to communicate securely with all units supporting or being supported by the sniper team.

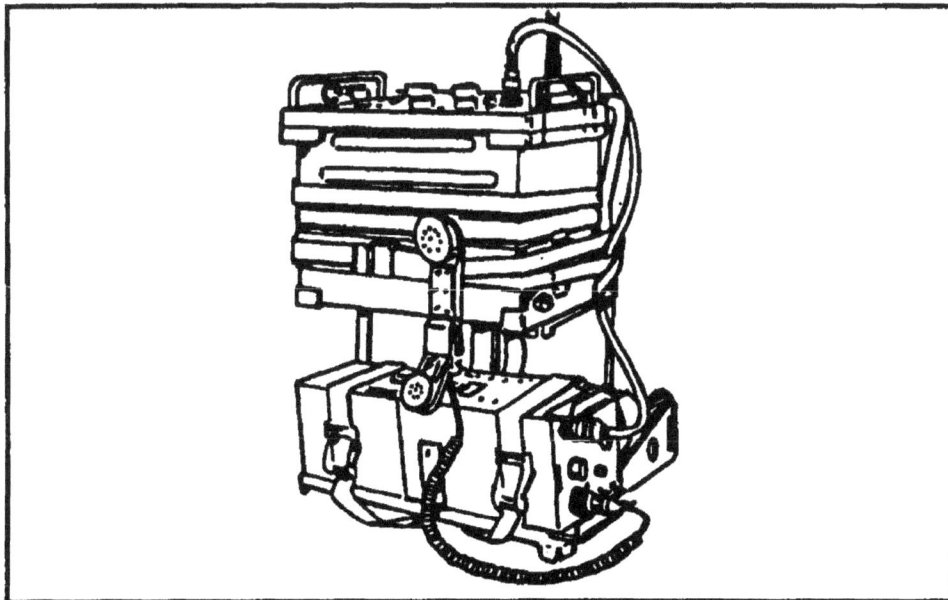

Figure 2-34. AN/PRC-77 radio.

2-19. AN/PRC-104A RADIO TRANSCEIVER

The AN/ PRC-104A is a state-of-the-am lightweight radio transceiver that operates in the high frequency and in the upper part of the low frequency portions of the radio spectrum (Figure 2-35). The receiver/ transmitter circuits can be tuned to any frequency between 2.0000 and 29.9999 MHz in 100 Hz increments, making it possible to tune up to 280,000 separate frequencies. The radio operates in the upper or lower side bank modes for voice communications, CW for Morse code, or FSK (frequency-shift keying) for transmission of teletype or other data.

a. In the man-pack configuration, the radio set is carried and operated by one man or, with the proper accessories, it can be configured for vehicle or fixed-station use. The radio set with antenna and handset weighs 15.7 pounds.

b. The control panel, human-engineered for ease of operation, makes it possible to adjust all controls even while wearing heavy gloves. Unlike older, similar radio sets, there are no front panel meters or indicator lights on the AN/ PRC-104A. All functions that formerly required these types of indicators are monitored by the radio and communicated to the operator as special tones in the handset. This feature is highly useful during tactical blackout operations. The superior design and innovative features of the AN/ PRC-104A radio set make it possible to maintain a reliable long-range communications link. The radio uses lightweight, portable equipment that can be operated by personnel who have minimum training.

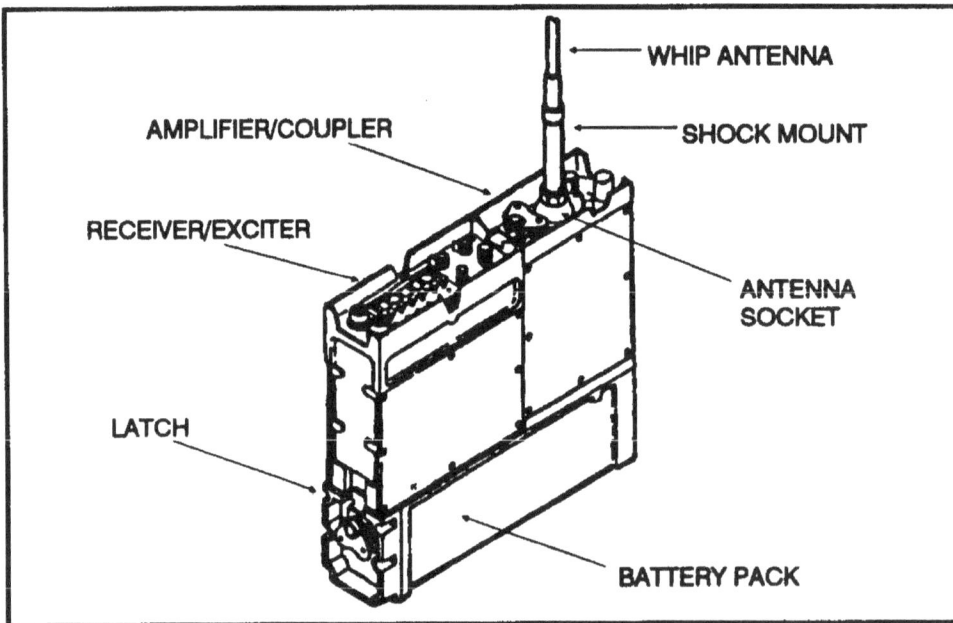

Figure 2-35. AN/PRC-104A radio transceiver.

2-20. AN/PRC-119 RADIO

The AN/ PRC-119 (Figure 2-36) replaces the AN/ PRC-77, although the AN/ PRC-77 is still in use. The AN/ PRC-l19 is a man-pack portable, VHF/ FM radio that is designed for simple, quick operation using a 16-element keypad for push-button tuning. It can also be used for short-range and long-range operation for voice, FSK, or digital data communications. It can also be used for single-channel operation or in a jam-resistant, frequency-hopping mode, which can be changed as needed This radio has a built-in self-test with visual and audio readbacks. It is compatible with the AN/ KY-57 for secure communications.

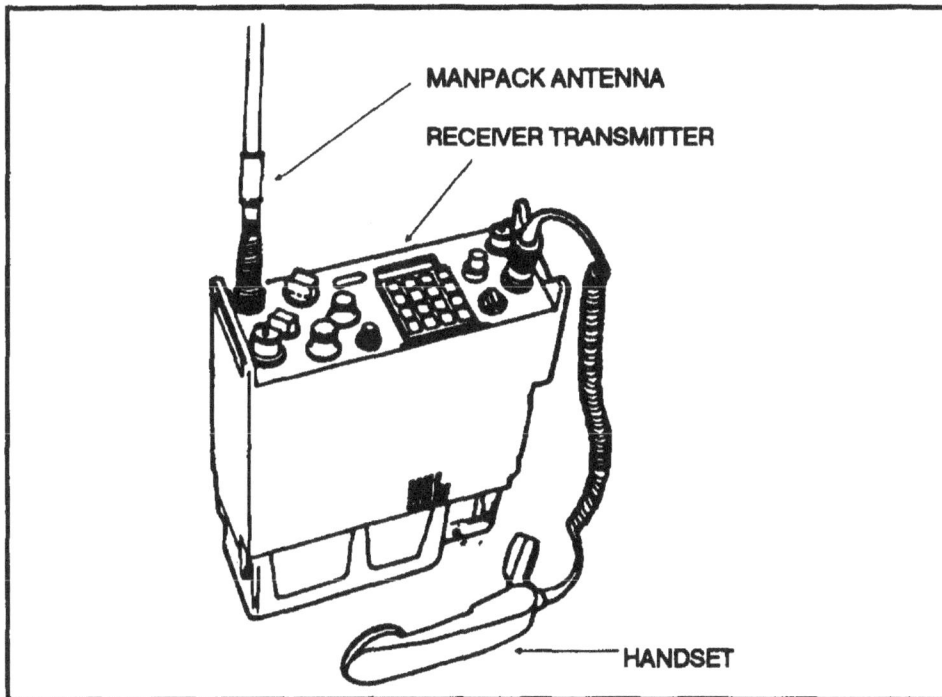

Figure 2-36. AN/PRC-119 radio.

CHAPTER 3

MARKSMANSHIP

*Sniper marksmanship is an extension of basic rifle marksmanship
and focuses on the techniques needed to engage targets at
extended ranges. To successfully engage targets at increased
distances, the sniper team must be proficient in marksmanship
fundamentals and advanced marksmanship skills. Examples of
these skills are determining the effects of weather conditions on
ballistics, holding off for elevation and windage, engaging moving
targets, using and adjusting scopes, and zeroing procedures.
Markmanship skills should be practiced often.*

Section I
FUNDAMENTALS

The sniper team must be thoroughly trained in the fundamentals
of marksmanship. These include assuming a position, aiming, breath
control, and trigger control. These fundamentals develop fixed and
correct firing habits for instinctive application. Every sniper should
periodically refamiliarize himself with these fundamentals regardless of
his experience.

3-1. STEADY POSITION ELEMENTS
The sniper should assume a good firing position (Figure 3-1, page 3-2) in
order to engage targets with any consistency. A good position enables the
sniper to relax and concentrate when preparing to fire.

a. **Position Elements.** Establishing a mental checklist of steady
position elements enhances the sniper's ability to achieve a first-round hit.

(1) *Nonfiring hand.* Use the nonfiring hand to support the butt of
the weapon. Place the hand next to the cheat and rest the tip of the butt
on it. Bail the hand into a fist to raise the weapon's butt or loosen the fist

to lower the weapon's butt. An effective method is to hold a sock full of sand in the nonfiring hand and to place the weapon butt on the sock. This reduces body contact with the weapon. To raise the butt, squeeze the sock and to lower it, loosen the grip on the sock.

(2) *Butt of the stock.* Place the butt of the stock firmly in the pocket of the shoulder. Insert a pad on the ghillie suit (see Chapter 4) where contact with the butt is made to reduce the effects of pulse beat and breathing, which can be transmitted to the weapon.

(3) *Firing hand.* With the firing hand, grip the small of the stock. Using the middle through little fingers, exert a slight rearward pull to keep the butt of the weapon firmly in the pocket of the shoulder. Place the thumb over the top of the small of the stock. Place the index finger on the trigger, ensuring it does not touch the stock of the weapon. This avoids disturbing the lay of the rifle when the trigger is squeezed.

(4) *Elbows.* Find a comfortable position that provides the greatest support.

(5) *Stock weld.* Place the cheek in the same place on the stock with each shot. A change in stock weld tends to cause poor sight alignment, reducing accuracy.

(6) *Bone support.* Bone support is the foundation of the firing position; they provide steady support of the weapon.

Figure 3-1. Firing position.

(7) *Muscle relaxation.* When using bone support, the sniper can relax muscles, reducing any movement that could be caused by tense or trembling muscles. Aside from tension in the trigger finger and firing hand, any use of the muscle generates movement of the sniper's cross hairs.

(8) *Natural point of aim.* The point at which the rifle naturally rest in relation to the aiming point is called natural point of aim.

(a) Once the sniper is in position and aimed in on his target, the method for checking for natural point of aim is for the sniper to close his eyes, take a couple of breaths, and relax as much as possible. Upon opening his eyes, the scope's cross hairs should be positioned at the sniper's preferred aiming point. Since the rifle becomes an extension of the sniper's body, it is necessary to adjust the position of the body until the rifle points naturally at the preferred aiming point on the target.

(b) Once the natural point of aim has been determined, the sniper must maintain his position to the target. To maintain his natural point of aim in all shooting positions, the natural point of aim can be readjusted and checked periodically.

(c) The sniper can change the elevation of the natural point of aim by leaving his elbows in place and by sliding his body forward or rearward. This raises or lowers the muzzle of the weapon, respectively. To maintain the natural point of aim after the weapon has been fired, proper bolt operation becomes critical. The sniper must practice reloading while in the prone position without removing the butt of the weapon from the firing shoulder. This may be difficult for the left-hand firer. The two techniques for accomplishing this task are as follows:

- After firing, move the bolt slowly to the rear while canting the weapon to the right. Execution of this task causes the spent cartridge to fall next to the weapon.
- After firing, move the bolt to the rear with the thumb of the firing hand. Using the index and middle fingers, reach into the receiver and catch the spent cartridge as it is being ejected. This technique does not require canting the weapon.

NOTE: The sniper conducts bolt operation under a veil or equivalent camouflage to improve concealment.

b. **Steady Firing Position.** On the battlefield, the sniper must assume a steady firing position with maximum use of cover and concealment. Considering the variables of terrain, vegetation, and tactical situations, the sniper can use many variations of the basic positions. When assuming a firing position, he must adhere to the following basic rules:

(1) Use any support available.

(2) Avoid touching the support with the barrel of the weapon since it interferes with barrel harmonics and reduces accuracy.

(3) Use a cushion between the weapon and the support to prevent slippage of the weapon.

(4) Use the prone supported position whenever possible.

c. **Types of Firing Positions.** Due to the importance of delivering precision fire, the sniper makes maximum use of artificial support and eliminates any variable that may prevent adhering to the basic rules. He uses the prone supported; prone unsupported; kneeling unsupported; kneeling, sling supported; standing supported; and the Hawkins firing positions.

(1) *Prone supported position.* The prone supported position is the steadiest position; it should be used whenever possible (Figure 3-2). To assume the prone supported position, the sniper should—

(a) Lie down and place the weapon on a support that allows pointing in the direction of the target. Keep the position as low as possible. (For field-expedient weapon supports, see paragraph 3-1d.)

(b) Remove the nonfiring hand from underneath the fore-end of the weapon by folding the arm underneath the receiver and trigger, grasping the rear sling swivel. This removes any chance of subconsciously trying to exert control over the weapon's natural point of aim. Keep the elbows in a comfortable position that provides the greatest support.

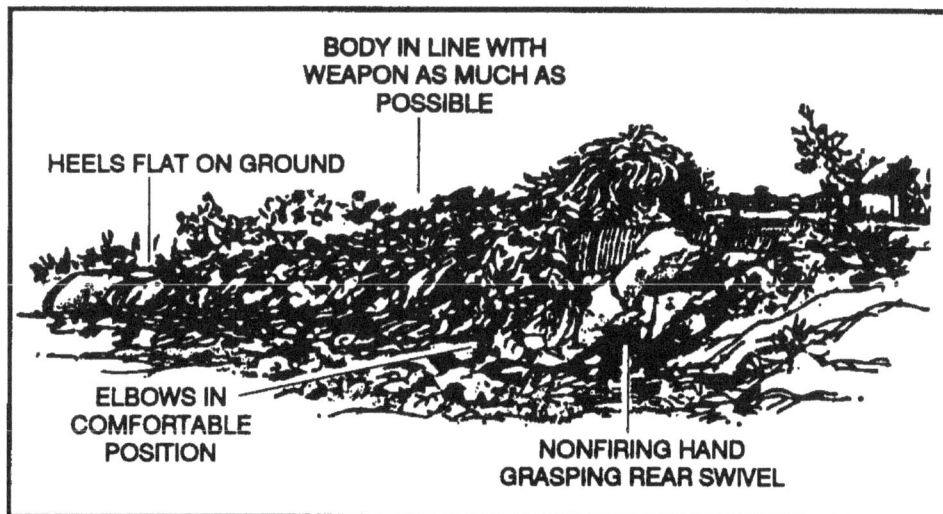

Figure 3-2. Prone supported position.

(c) Keep the body in line with the weapon as much as possible-not at an angle. This presents less of a target to the enemy and more body mass to absorb recoil.

(d) Spread legs a comfortable distance apart with the heels on the ground or as close as possible without causing strain.

(2) *Prone unsupported position.* The prone unsupported position (Figure 3-3) offers another stable firing platform for engaging targets. To assume this position, the sniper faces his target, spreads his feet a comfortable distance apart, and drops to his knees. Using the butt of the rifle as a pivot, the firer rolls onto his nonfiring side. He places the rifle butt in the pocket formed by the firing shoulder, grasps the pistol grip in his firing hand, and lowers the firing elbow to the ground. The rifle rests in the V formed by the thumb and fingers of the nonfiring hand The sniper adjusts the position of his firing elbow until his shoulders are about level, and pulls back firmly on the rifle with both hands. To complete the position, he obtains a stock weld and relaxes, keeping his heels close to the ground.

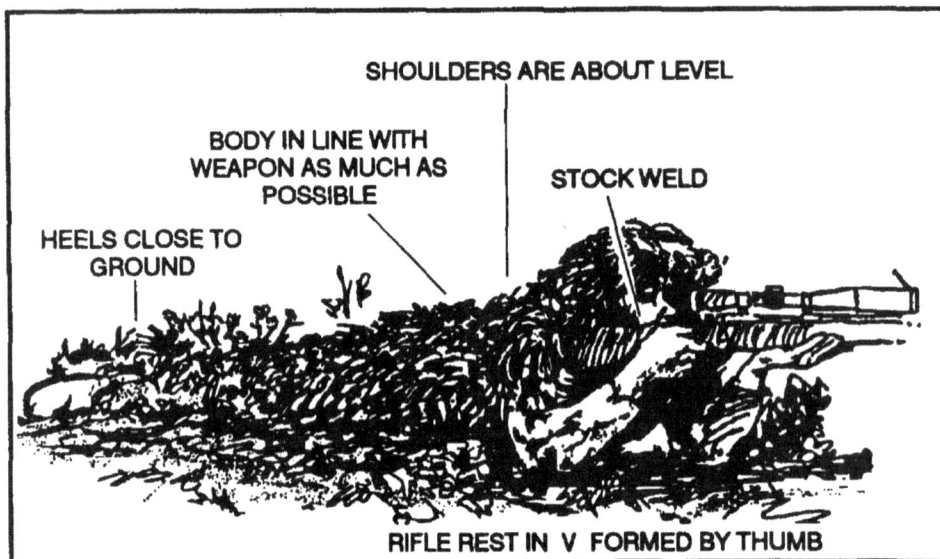

Figure 3-3. Prone unsupported position.

(3) *Kneeling unsupported position.* The kneeling unsupported position (Figure 3-4, page 3-6) is assumed quickly. It places the sniper high enough to see over small brush and provides for a stable position.

(a) Place the body at a 45-degree angle to the target.

(b) Kneel and place the right knee on the ground.

(c) Keep the left leg as perpendicular to the ground as possible; sit back on the right heel, placing it as directly under the spinal column as possible. A variation is to turn the toe inward and sit squarely on the right foot.

(d) Grasp the small of the stock of the weapon with the firing hand, and cradle the fore-end of the weapon in a crook formed with the left arm.

(e) Place the butt of the weapon in the pocket of the shoulder, then place the meaty underside of the left elbow on top of the left knee.

(f) Reach under the weapon with the left hand, and lightly grasp the firing arm.

(g) Relax forward and into the support position, using the left shoulder as a contact point. This reduces transmission of the pulsebeat into the sight picture.

(h) Lean against a tree, building, or vehicle for body support.

Figure 3-4. Kneeling unsupported position.

(4) *Kneeling, sling supported position.* If vegetation presents a problem, the sniper can raise his kneeling position by using the rifle sling. To assume the kneeling, sling supported position, he executes the first three steps for assuming a kneeling unsupported position. With the leather sling mounted to the weapon, the sniper turns the sling one-quarter turn to the left. The lower part of the sling will then form a loop.

(a) Place the left arm (nonfiring) through the loop; pull the sling up the arm and place it on the upper arm between the elbow and shoulder, but not directly over the biceps.

(b) Tighten the sling by sliding the sling keeper against the loop holding the arm.

(c) Rotate the left arm in a clockwise motion around the sling and under the rifle with the sling secured to the upper arm. Place the fore-end of the stock in the V formed by the thumb and forefinger of the left hand. Relax the left arm and hand, let the sling support the weight of the weapon.

(d) Place the butt of the rifle against the right shoulder and place the left elbow on top of the left knee (Figure 3-5). Pull the left hand back along the fore-end of the rifle toward the trigger guard to add to stability.

Figure 3-5. Kneeling, sling supported position.

(5) *Standing supported position.* The standing supported position is the least steady of the supported positions and should be used only as a last resort (Figure 3-6, page 3-8).

(a) To assume the standing supported position with horizontal support, such as a wall or ledge, the sniper proceeds as follows:

• Locate a solid object for support. Avoid branches as they tend to sway when wind is present.

- Form a V with the thumb and forefinger of the nonfiring hand.
- Place the nonfiring hand against the support with the fore-end of the weapon resting in the V of the hand. This steadies the weapon and allows quick recovery from recoil.
- Then place the butt of the weapon in the pocket of the shoulder.

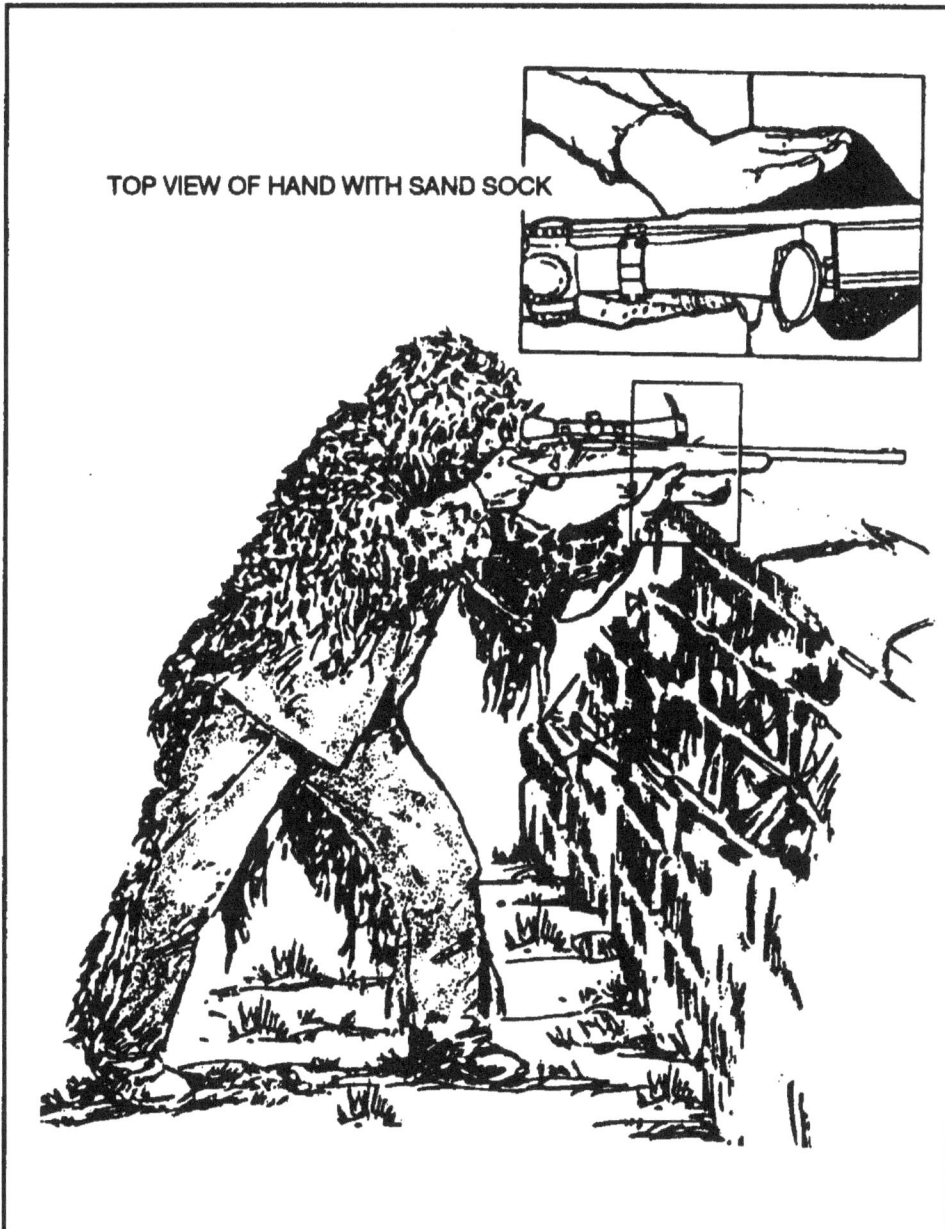

TOP VIEW OF HAND WITH SAND SOCK

**Figure 3-6. Standing supported position
(horizontal support).**

(b) To use vertical support (Figure 3-7), such as a tree, telephone pole, corner of building, or vehicle, the sniper proceeds as follows:

- Locate stable support. Face the target, then turn 45 degrees to the right of the target, and place the palm of the nonfiring hand at arm's length against the support.
- Lock the left arm straight, let the left leg buckle, and place body weight against the nonfiring hand. Keep the trail leg straight.
- Place the fore-end of the weapon in the V formed by extending the thumb of the nonfiring hand.
- Exert more pressure to the rear with the firing hand.

Figure 3-7. Standing supported position (vertical support).

(6) *Hawkins position.* The Hawkins position (Figure 3-8) is a variation of the prone unsupported position. The sniper uses it when firing from a low bank or a depression in the ground, over a roof, or so forth. It cannot be used on level ground since the muzzle cannot be raised high enough to aim at the target. It is a low-profile position with excellent stability and aids concealment. To assume this position, the sniper uses the weapon's sling and proceeds as follows:

CAUTION
LOCK THE NONFIRING ARM STRAIGHT OR THE FACE WILL ABSORB THE WEAPON'S RECOIL.

(a) After assuming a prone position, grasp the upper sling swivel and sling with the nonfiring hand, forming a fist to support the front of the weapon.

(b) Ensure the nonfiring arm is locked straight since it will absorb the weapon's recoil. Wearing a glove is advisable.

(c) Rest the butt of the weapon on the ground and place it *under* the firing shoulder.

The sniper can make minor adjustments in muzzle elevation by tightening or relaxing the fist of the nonfiring hand. If more elevation is required, he can place a support under the nonfiring fist.

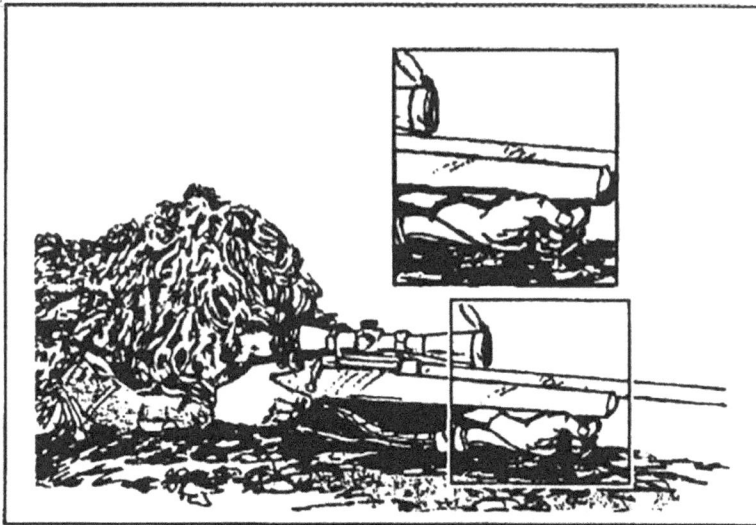

Figure 3-8. Hawkins position.

d. **Field-Expedient Weapon Support.** Support of the weapon is critical to the sniper's success in engaging targets. Unlike a well-equipped firing range with sandbags for weapon support, the sniper can encounter situations where weapon support relies on common sense and imagination. The sniper should practice using these supports at every opportunity and select the one that best suits his needs. He must train as if in combat to avoid confusion and self-doubt. The following items are commonly used as field-expedient weapon supports

(1) *Sand sock.* The sniper needs the sand sock when delivering precision fire at long ranges. He uses a standard issue, olive-drab wool sock filled one-half to three-quarters full of sand and knotted off. He places it under the rear sling swivel when in the prone supported position for added stability (Figure 3-9). By limiting minor movement and reducing pulse beat, the sniper can concentrate on trigger control and aiming. He uses the nonfiring hand to grip the sand sock, rather than the rear sling swivel. The sniper makes minor changes in muzzle elevation by squeezing or relaxing his grip on the sock. He uses the sand sock as padding between the weapon and a rigid support also.

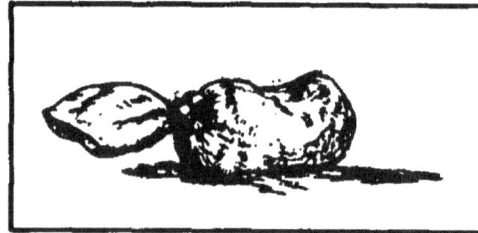

Figure 3-9. Sand sock.

(2) *Rucksack.* If the sniper is in terrain without any natural support, he may use his rucksack (Figure 3-10). He must consider the height and presence of rigid objects within the rucksack. The rucksack must conform to weapon contours to add stability.

Figure 3-10. Rucksack.

(3) *Sandbag.* The sniper can fill an empty sandbag (Figure 3-11) on site.

Figure 3-11. Sandbag.

(4) *Tripod.* The sniper can build a field-expedient tripod (Figure 3-12) by tying together three 12-inch long sticks (one thicker than the others) with 550 cord or the equivalent. When tying the sticks, he wraps the cord at the center point and leaves enough slack to fold the legs out into a triangular base. Then, he places the fore-end of the weapon between the three uprights.

(5) *Bipod.* The sniper can build a field-expedient bipod (Figure 3-12) by tying together two 12-inch sticks, thick enough to support the weight of the weapon. Using 550 cord or the equivalent, he ties the sticks at the center point, leaving enough slack to fold them out in a scissor-like manner. He then places the weapon between the two uprights. The bipod is not as stable as other field-expedient items, and it should be used only in the absence of other techniques.

(6) *Forked stake.* The tactical situation determines the use of the forked stake. Unless the sniper can drive a forked stake into the ground, this is the least desirable of the techniques; that is, he must use his nonfiring hand to hold the stake in an upright position (Figure 3-12). Delivering long-range precision fire is a near-impossibility due to the unsteadiness of the position.

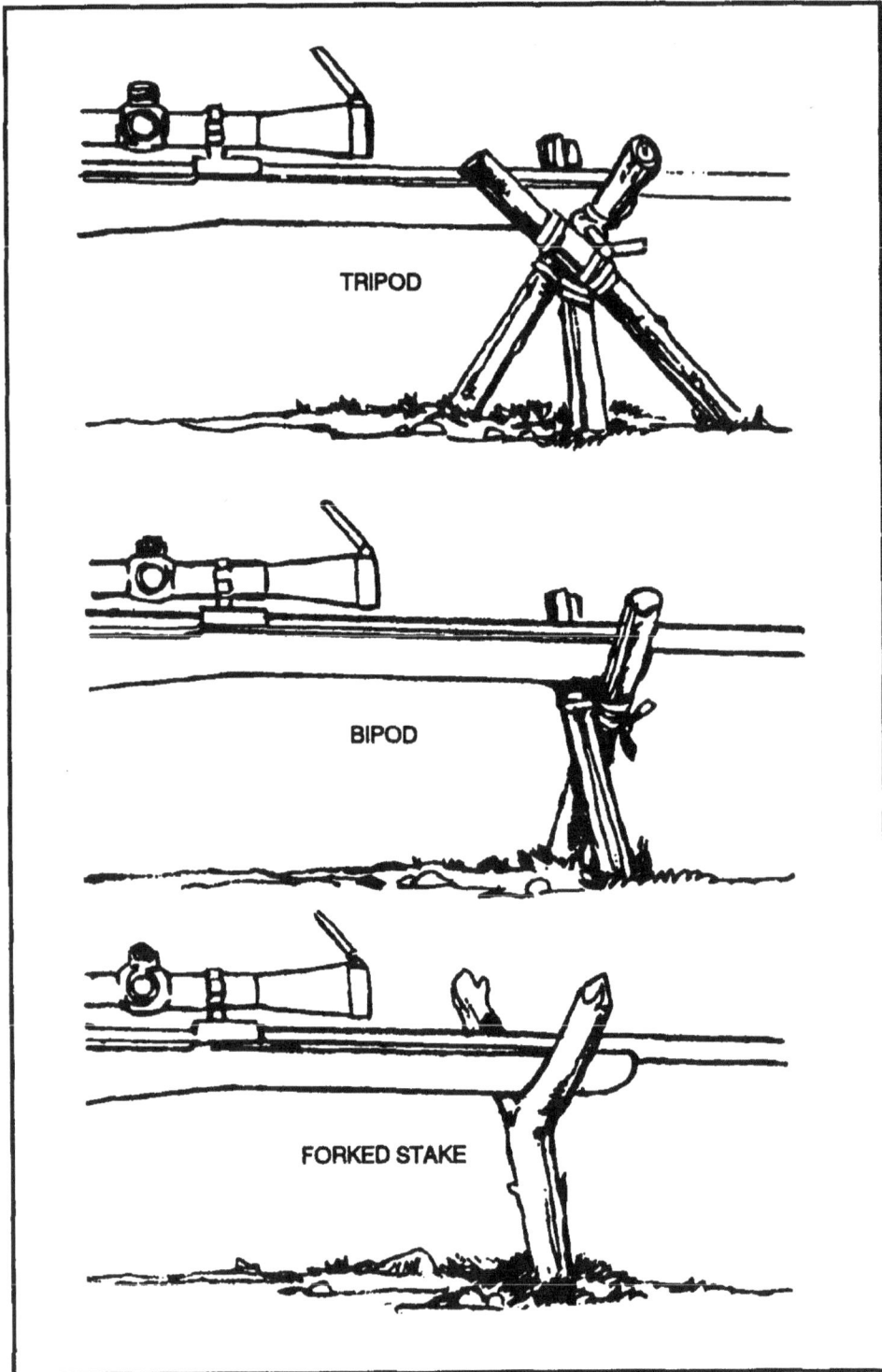

Figure 3-12. Field-expedient tripod, bipod, and forked stake.

e. **Sniper and Observer Positioning.** The sniper should find a place on the ground that allows him to build a steady, comfortable position with the best cover, concealment, and visibility of the target area. Once established, the observer should position himself out of the sniper's field of view on his firing side.

(1) The closer the observer gets his spotting telescope to the sniper's line of bore, the easier it is to follow the trace (path) of the bullet and observe the point of impact. A position at 4 to 5 o'clock (7 to 8 o'clock for left-handed firers) from the firing shoulder and close to (but not touching) the sniper is best (Figure 3-13).

NOTE: Trace is the visible trail of a bullet and is created by the shock wave of a supersonic bullet. The shockwave compresses the air along the leading edge of a bullet causing water vapor in the air to momentary condense and become visible. To the observer, located to the rear of the sniper, trace appears as a rapidly moving V-shaped vortex in the air following the trajectory of the bullet. Through close observation and practice, trace can be used to judge the bullet's trajectory relative to the aiming point, making corrections easier for a follow-up shot. Trace can best be seen if the observer's optics are directly in line with the axis of the sniper's rifle barrel. Watching the trace and the effects of the bullet's impact are the primary means by which the observer assists the sniper in calling the shot.

Figure 3-13. Sniper team positioning.

(2) If the sniper is without weapon support in his position, he uses the observer's body as a support (Figure 3-14). This support is not recommended since the sniper must contend with his own movement and the observer's body movement. The sniper should practice and prepare to use an observer supported position. A variety of positions can be used; however, the two most stable are when the observer is in a prone or sitting position.

(a) *Prone.* To assume the prone position, the observer lies at a 45-to 75-degree angle to the target and observes the area through his spotting telescope. The sniper assumes a a prone supported position, using the back of the observer's thigh for support. Due to the offset angle, the observer may only see the bullet impact.

Figure 3-14. Prone observer supported position.

(b) *Sitting.* If vegetation prevents the sniper from assuming a prone position, the sniper has the observer face the target area and assume a cross-legged sitting position. The observer places his elbows on his knees to stabilize his position. For observation, the observer uses binoculars held in his hands. The spotting telescope is not recommended due to its higher magnification and the unsteadiness of this position. The sniper is behind the observer in an open-legged, cross-legged, or kneeling position, depending on the target's elevation (Figure 3-15, page 3-16). The sniper places the fore-end of the weapon across the observer's left shoulder, stabilizing the weapon with the forefinger of the nonfiring hand. When using these positions, the sniper's effective engagement of targets at extended ranges is difficult and used only as a last resort. When practicing these positions, the sniper and observer must enter respiratory pause together to eliminate movement from breathing.

WEAPON PLACEMENT IS ON THE
OBSERVER'S LEFT SHOULDER . . .
OPEN-LEGGED, CROSS-LEGGED,
OR KNEELING POSITION

NONFIRING HAND

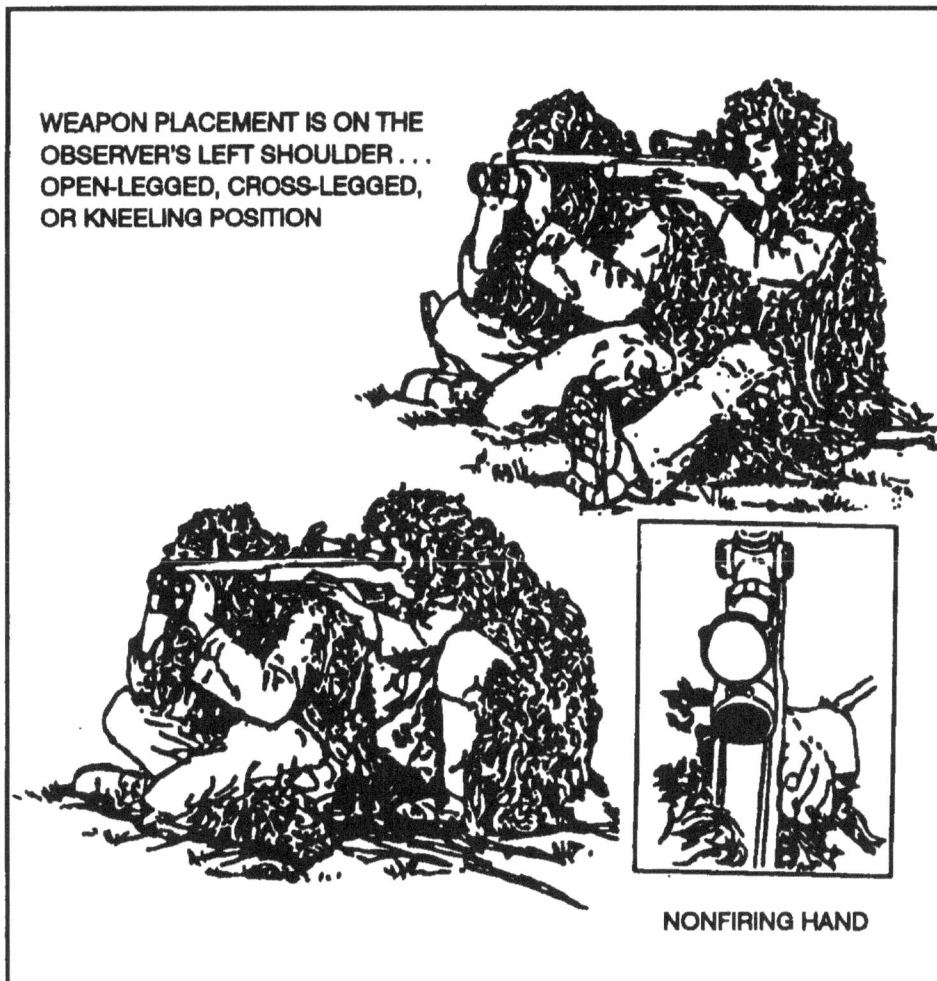

Figure 3-15. Sitting position.

3-2. AIMING

The sniper begins the aiming process by aligning the rifle with the target
when assuming a firing position. He should point the rifle naturally at
the desired point of aim. If his muscles are used to adjust the weapon onto
the point of aim, they automatically relax as the rifle fires, and the rifle
begins to move toward its natural point of aim. Because this movement
begins just before the weapon discharge, the rifle is moving as the bullet
leaves the muzzle. This causes inaccurate shots with no apparent cause
(recoil disguises the movement). By adjusting the weapon and body as a
single unit, rechecking, and readjusting as needed, the sniper achieves a
true natural point of aim. Once the position is established, the sniper
then aims the weapon at the exact point on the target. Aiming involves:
eye relief, sight alignment, and sight picture.

a. **Eye Relief.** This is the distance from the sniper's firing eye to the rear sight or the rear of the scope tube. When using iron sights, the sniper ensures the distance remains consistent from shot to shot to preclude changing what he views through the rear sight. However, relief will vary from firing position to firing position and from sniper to sniper, according to the sniper's neck length, his angle of head approach to the stock, the depth of his shoulder pocket, and his firing position. This distance (Figure 3-16) is more rigidly controlled with telescopic sights than with iron sights. The sniper must take care to prevent eye injury caused by the scope tube striking his brow during recoil. Regardless of the sighting system he uses, he must place his head as upright as possible with his firing eye located directly behind the rear portion of the sighting system. This head placement also allows the muscles surrounding his eye to relax. Incorrect head placement causes the sniper to look out of the top or corner of his eye, resulting in muscular strain. Such strain leads to blurred vision and can also cause eye strain. The sniper can avoid eye strain by not staring through the telescopic or iron sights for extended periods. The best aid to consistent eye relief is maintaining the same stock weld from shot to shot.

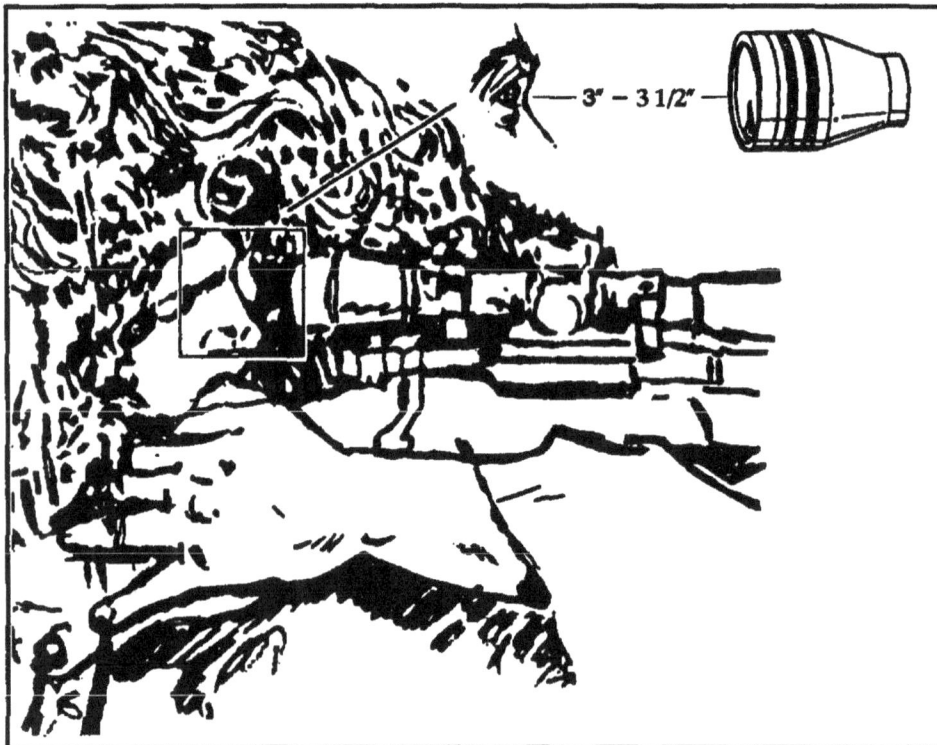

Figure 3-16. Eye relief.

b. **Sight Alignment.** With telescopic sights, sight alignment is the relationship between the cross hairs (reticle) and a full field of view as seen by the sniper. The sniper must place his head so that a full field of view fills the tube, with no dark shadows or crescents to cause inaccurate shots. He centers the reticle in a full field of view, ensuring the vertical cross hair is straight up and down so the rifle is not canted. Again, the center is easiest for the sniper to locate and allows for consistent reticle placement. With iron sights, sight alignment is the relationship between the front and rear sights as seen by the sniper (Figure 3-17). The sniper centers the top edge of the front sight blade horizontally and vertically within the rear aperture. (The center of aperture is easiest for the eye to locate and allows the sniper to be consistent in blade location.)

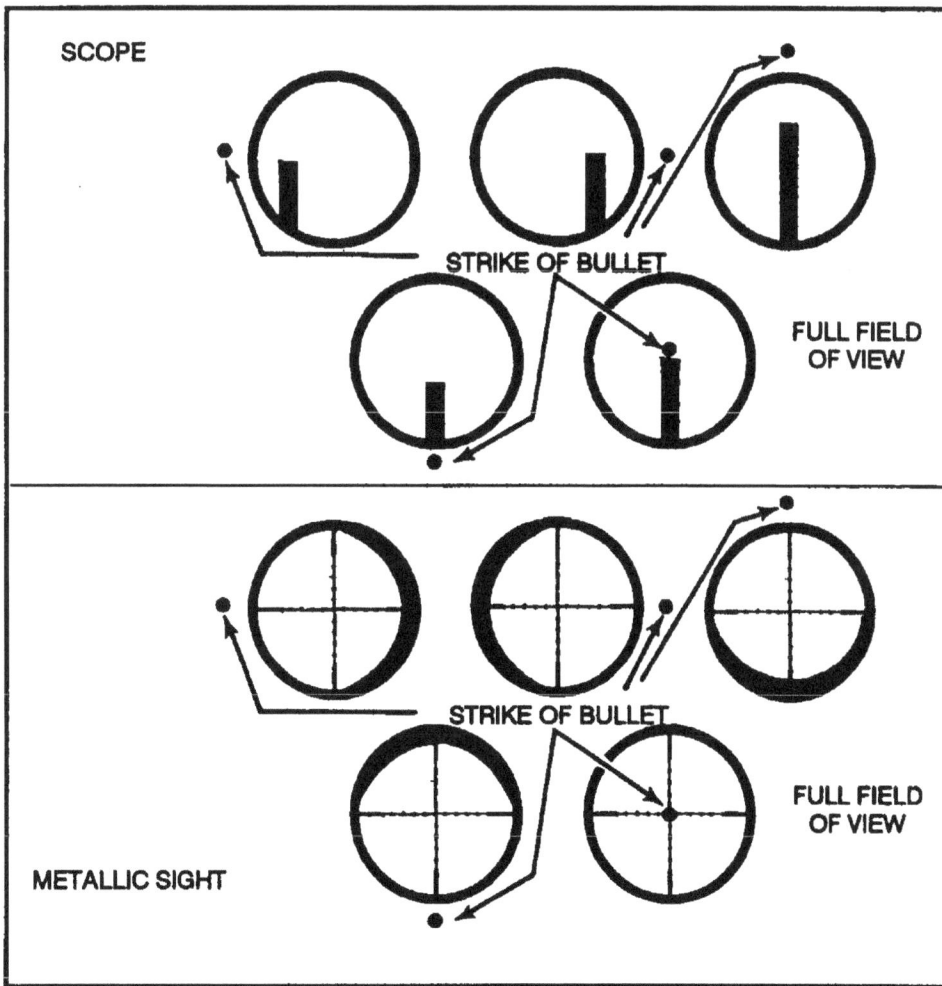

Figure 3-17. Sight alignment.

e. **Sight Picture.** With telescopic sights, the sight picture is the relationship between the reticle and full field of view and the target as seen by the sniper. The sniper centers the reticle in a full field of view. He then places the reticle center of the largest visible mass of the target (as in iron sights). The center of mass of the target is easiest for the sniper to locate, and it surrounds the intended point of impact with a maximum amount of target area. With iron sights, sight picture is the relationship between the rear aperture, the front sight blade, and the target as seen by the sniper (Figure 3-18). The sniper centers the top edge of the blade in the rear aperture. He then places the top edge of the blade in the center of the largest visible mass of the target (disregard the head and use the center of the torso).

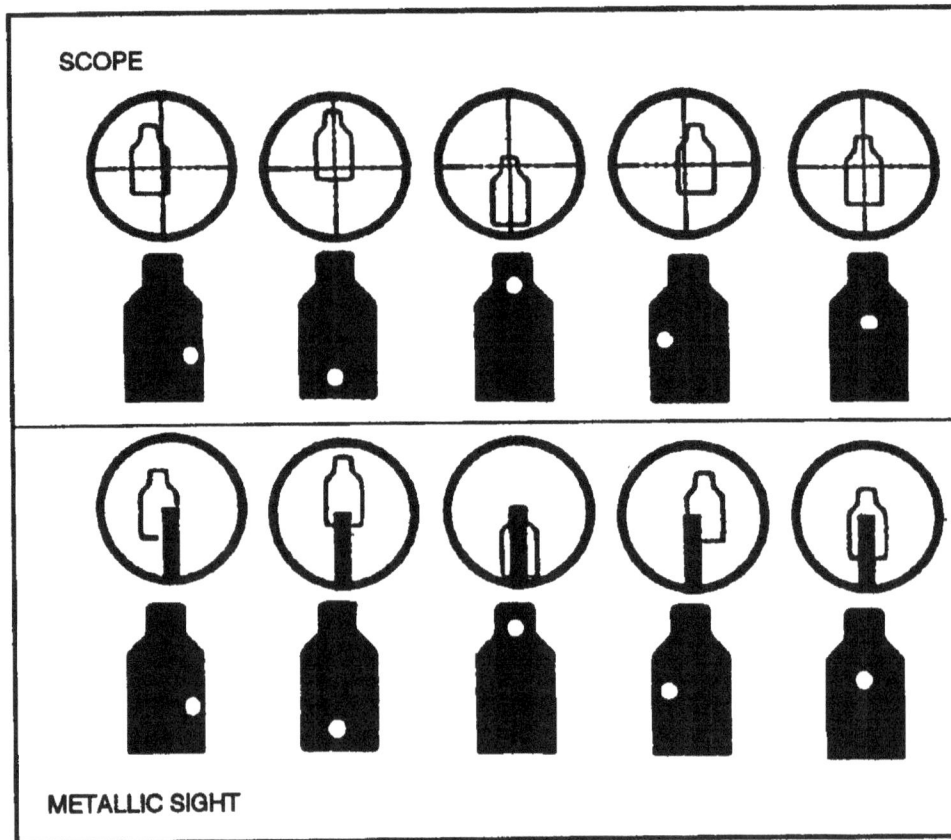

Figure 3-18. Sight picture.

d. **Sight Alignment Error.** When sight alignment and picture are perfect (regardless of sighting system) and all else is done correctly, the shot will hit center of mass on the target. However, with an error insight alignment, the bullet is displaced in the direction of the error. Such an error creates an angular displacement between the line of sight and the line of bore. This displacement increases as range increases; the amount of bullet displacement depends on the size of alignment error. Close targets show little or no visible error. Distant targets can show great displacement or can be missed altogether due to severe sight misalignment. An inexperienced sniper is prone to this kind of error, since he is unsure of what correctly aligned sights look like (especially telescopic sights); a sniper varies his head position (and eye relief) from shot to shot, and he is apt to make mistakes while firing.

e. **Sight Picture Error.** An error in sight picture is an error in the placement of the aiming point. This causes no displacement between the line of sight and the line of bore. The weapon is simply pointed at the wrong spot on the target. Because no displacement exists as range increases, close and far targets are hit or missed depending on where the front sight or the reticle is when the rifle fires. All snipers face this kind of error every time they shoot. This is because, regardless of firing position stability, the weapon will always be moving. A supported rifle moves much leas than an unsupported one, but both still move in what is known as a *wobble area.* The sniper must adjust his firing position so that his wobble area is as small as possible and centered on the target. With proper adjustments, the sniper should be able to fire the shot while the front sight blade or reticle is on the target at, or very near, the desired aiming point. How far the blade or reticle is from this point when the weapon fires is the amount of sight picture error all snipers face.

f. **Dominant Eye.** To determine which eye is dominant, the sniper extends one arm to the front and points the index finger skyward to select an aiming point. With both eyes open, he aligns the index finger with the aiming point, then closes one eye at a time while looking at the aiming point. One eye will make the finger appear to move off the aiming point; the other eye will stay on the aiming point. The dominant eye is the eye that does not move the finger from the aiming point. Some individuals may have difficulty aiming because of interference from their dominant eye, if this is not the eye used in the aiming process. This may require the sniper to fire from the other side of the weapon (right-handed firer will fire left-handed). Such individuals must close the dominant eye while shooting.

3-3. BREATH CONTROL

Breath control is important with respect to the aiming process. If the sniper breathes while trying to aim, the rise and fall of his chest causes the rifle to move. He must, therefore, accomplish sight alignment during breathing. To do this, he first inhales then exhales normally and stops at the moment of natural respiratory pause.

a. A respiratory cycle lasts 4 to 5 seconds. Inhalation and exhalation require only about 2 seconds. Thus, between each respiratory cycle there is a pause of 2 to 3 seconds. This pause can be extended to 10 seconds without any special effort or unpleasant sensations. The sniper should shoot during this pause when his breathing muscles relax. This avoids strain on his diaphragm.

b. A sniper should assume his firing position and breathe naturally until his hold begins to settle. Many snipers then take a slightly deeper breath, exhale, and pause, expecting to fire the shot during the pause. If the hold does not settle enough to allow the shot to be fired, the sniper resumes normal breathing and repeats the process.

c. The respiratory pause should never feel unnatural. If it is too long, the body suffers from oxygen deficiency and sends out signals to resume breathing. These signals produce involuntary movements in the diaphragm and interfere with the sniper's ability to concentrate. About 8 to 10 seconds is the maximum safe period for the respiratory pause. During multiple, rapid engagements, the breathing cycle should be forced through a rapid, shallow cycle between shots instead of trying to hold the breath or breathing. Firing should be accomplished at the forced respiratory pause.

3-4. TRIGGER CONTROL

Trigger control is the most important of the sniper marksmanship fundamentals. It is defined as causing the rifle to fire when the sight picture is at its best, without causing the rifle to move. Trigger squeeze is uniformly increasing pressure straight to the rear until the rifle fires.

a. Proper trigger control occurs when the sniper places his firing finger as low on the trigger as possible and still clears the trigger guard, thereby achieving maximum mechanical advantage and movement of the finger to the entire rifle.

b. The sniper maintains trigger control beat by assuming a stable position, adjusting on the target, and beginning a breathing cycle. As the sniper exhales the final breath toward a natural respiratory pause, he secures his finger on the trigger. As the front blade or reticle settles at the desired point of aim, and the natural respiratory pause is entered, the sniper applies initial pressure. He increases the tension on the trigger during the respiratory pause as long as the front blade or reticle remains in the area of the target that ensures a well-placed shot. If the front blade or reticle moves away from the desired point of aim on the target, and the pause is free of strain or tension, the sniper stops increasing the tension on the trigger, waits for the front blade or reticle to return to the desired point, and then continues to squeeze the trigger. If movement is too large for recovery or if the pause has become uncomfortable (extended too long), the sniper should carefully release the pressure on the trigger and begin the respiratory cycle again.

c. As the stability of a firing position decreases, the wobble area increases. The larger the wobble area, the harder it is to fire the shot without reacting to it. This reaction occurs when the sniper—

(1) *Anticipates recoil.* The firing shoulder begins to move forward just before the round fires.

(2) *Jerks the trigger.* The trigger finger moves the trigger in a quick, choppy, spasmodic attempt to fire the shot before the front blade or reticle can move away from the desired point of aim.

(3) *Flinches.* The sniper's entire upper body (or parts thereof) overreacts to anticipated noise or recoil. This is usually due to unfamiliarity with the weapon.

(4) *Avoids recoil.* The sniper tries to avoid recoil or noise by moving away from the weapon or by closing the firing eye just before the round fires. This, again, is caused by a lack of knowledge of the weapon's actions upon firing.

3-5. FOLLOW-THROUGH

Applying the fundamentals increases the odds of a well-aimed shot being fired. When mastered, additional skills can make that first-round kill even more of a certainty. One of these skills is the follow-through.

a. Follow-through is the act of continuing to apply all the sniper marksmanship fundamentals as the weapon fires as well as immediately after it fires. It consists of—

(1) Keeping the head infirm contact with the stock (stock weld).

(2) Keeping the finger on the trigger all the way to the rear.

(3) Continuing to look through the rear aperture or scope tube.

(4) Keeping muscles relaxed.

(5) Avoiding reaction to recoil and or noise.

(6) Releasing the trigger only after the recoil has stopped.

b. A good follow-through ensures the weapon is allowed to fire and recoil naturally. The sniper/ rifle combination reacts as a single unit to such actions.

3-6. CALLING THE SHOT

Calling the shot is being able to tell where the round should impact on the target. Because live targets invariably move when hit, the sniper will find it almost impossible to use his scope to locate the target after the round is fired. Using iron sights, the sniper will find that searching for a downrange hit is beyond his abilities. He must be able to accurately call his shots. Proper follow-through will aid in calling the shot. The dominant factor in shot calling is knowing where the reticle or blade is located when the weapon discharges. This location is called the *final focus point*.

a. With iron sights, the final focus point should be on the top edge of the front sight blade. The blade is the only part of the sight picture that is moving (in the wobble area). Focusing on it aids in calling the shot and detecting any errors insight alignment or sight picture. Of course, lining up the sights and the target initially requires the sniper to shift his focus from the target to the blade and back until he is satisfied that he is properly aligned with the target. This shifting exposes two more facts about eye focus. The eye can instantly shift focus from near objects (the blade) to far objects (the target).

b. The final focus is easily placed with telescopic sights because of the sight's optical qualities. Properly focused, a scope should present both the field of view and the reticle in sharp detail. Final focus should then be on the target. While focusing on the target, the sniper moves his head slightly from side to side. The reticle may seem to move across the target face, even though the rifle and scope are motionless. This movement is *parallax*. Parallax is present when the target image is not correctly focused on the reticle's focal plane. Therefore, the target image and the reticle appear to be in two separate positions inside the scope, causing the effect of reticle movement across the target. The M3A scope on the M24 has a focus adjustment that eliminates parallax in the scope. The sniper should adjust the focus knob until the target's image is on the same focal plane as the reticle. To determine if the target's image appears at the ideal location, the sniper should move his head slightly left and right to see if the reticle appears to move. If it does not move, the focus is properly adjusted and no parallax will be present.

3-7. INTEGRATED ACT OF FIRING

Once the sniper has been taught the fundamentals of marksmanship, his primary concern is his ability to apply it in the performance of his mission. An effective method of applying fundamentals is through the use of the integrated act of firing one round. The integrated act is a logical, step-by-step development of fundamentals whereby the sniper can develop habits that enable him to fire each shot the same way. The integrated act of firing can be divided into four distinct phases:

a. Preparation Phase. Before departing the preparation area, the sniper ensures that—

(1) The team is mentally conditioned and knows what mission they are to accomplish.

(2) A systematic check is made of equipment for completeness and serviceability including, but not limited to—

(a) Properly cleaned and lubricated rifles.

(b) Properly mounted and torqued scopes.

(c) Zero-sighted systems and recorded data in the sniper data book.

(d) Study of the weather conditions to determine their possible effects on the team's performance of the mission.

b. **Before-Firing Phase.** On arrival at the mission site, the team exercises care in selecting positions. The sniper ensures the selected positions support the mission. During this phase, the sniper—

(1) Maintains strict adherence to the fundamentals of position. He ensures that the firing position is as relaxed as possible, making the most of available external support. He also makes sure the support is stable, conforms to the position, and allows a correct, natural point of aim for each designated area or target.

(2) Once in position, removes the scope covers and checks the field(s) of fire, making any needed corrections to ensure clear, unobstructed firing lanes.

(3) Makes dry firing and natural point of aim checks.

(4) Double-checks ammunition for serviceability and completes final magazine loading.

(5) Notifies the observer he is ready to engage targets. The observer must be constantly aware of weather conditions that may affect the accuracy of the shots. He must also stay ahead of the tactical situation.

c. **Firing Phase.** Upon detection, or if directed to a suitable target, the sniper makes appropriate sight changes, aims, and tells the observer he is ready to fire. The observer then gives the needed windage and observes the target. To fire the rifle, the sniper should remember the key word, "BRASS." Each letter is explained as follows:

(1) *Breathe.* The sniper inhales and exhales to the natural respiratory pause. He checks for consistent head placement and stock weld. He ensures eye relief is correct (full field of view through the scope; no shadows present). At the same time, he begins aligning the cross hairs or front blade with the target at the desired point of aim.

(2) *Relax.* As the sniper exhales, he relaxes as many muscles as possible, while maintaining control of the weapon and position.

(3) *Aim.* If the sniper has a good, natural point of aim, the rifle points at the desired target during the respiratory pause. If the aim is off, the sniper should make a slight adjustment to acquire the desired point of aim. He avoids "muscling" the weapon toward the aiming point.

(4) **Squeeze.** As long as the sight picture is satisfactory, the sniper squeezes the trigger. The pressure applied to the trigger must be straight to the rear without disturbing the lay of the rifle or the desired point of aim.

d. After-Firing Phase. The sniper must analyze his performance If the shot impacted at the desired spot (a target hit), it may be assumed the integrated act of firing one round was correctly followed. If however, the shot was off call, the sniper and observer must check for Possible errors.

(1) Failure to follow the keyword, BRASS (partial field of view, breath held incorrectly, trigger jerked, rifle muscled into position, and so on).

(2) Target improperly ranged with scope (causing high or low shots).

(3) Incorrectly compensated for wind (causing right or left shots).

(4) Possible weapon/ammunition malfunction (used only as a last resort when no other errors are detected).

Once the probable reasons for an off-call shot is determined the sniper must make note of the errors. He pays close attention to the problem areas to increase the accuracy of future shots.

Section II
BALLISTICS

As applied to sniper marksmanship, types of ballistics may be defined as the study of the firing, flight, and effect of ammunition. Proper execution of marksmanship fundamentals and a thorough knowledge of ballistics ensure the successful completion of the mission. Tables and formulas in this section should be used only as guidelines since every rifle performs differently. Maximum ballistics data eventually result in a well-kept sniper data book and knowledge gained through experience.

3-8. TYPES OF BALLISTICS
Ballistics are divided into three distinct types: internal external, and terminal.

a. Internal-the interior workings of a weapon and the functioning of its ammunition.

b. External-the flight of the bullet from the muzzle to the target.

c. Termninal-what happens to the bullet after it hits the target. (See paragraph 3-16.)

3-9. TERMINOLOGY
To fully understand ballistics, the sniper should be familiar with the following terms:

a. Muzzle Velocity-the speed of the bullet as it leaves the rifle barrel, measured in feet per second. It varies according to various factors, such as ammunition type and lot number, temperature, and humidity.

b. Line of Sight- straight line from the eye through the aiming device to the point of aim.

c. Line of Departure-the line defined by the bore of the rifle or the path the bullet would take without gravity.

d. Trajectory-the path of the bullet as it travels to the target.

e. Midrange Trajectory/ Maximum Ordinate-the highest point the bullet reaches on its way to the target. This point must be known to engage a target that requires firing underneath an overhead obstacle, such as a bridge or a tree. In attention to midrange trajectory may cause the sniper to hit the obstacle instead of the target.

f. Bullet Drop—how far the bullet drops from the line of departure to the point of impact.

g. Time of Flight-the amount of time it takes for the bullet to reach the target from the time the round exits the rifle.

h. Retained Velocity-the speed of the bullet when it reaches the target. Due to drag, the velocity will be reduced.

3-10. EFFECTS ON TRAJECTORY

To be effective, the sniper must know marksmanship fundamentals and what effect gravity and drag will have on those fundamentals.

a. **Gravity.** As soon as the bullet exits the muzzle of the weapon, gravity begins to pull it down, requiring the sniper to use his elevation adjustment. At extended ranges, the sniper actually aims the muzzle of his rifle above his line of sight and lets gravity pull the bullet down into the target. Gravity is always present, and the sniper must compensate for this through elevation adjustments or hold-off techniques.

b. **Drag.** Drag is the slowing effect the atmosphere has on the bullet. This effect decreases the speed of the bullet according to the air—that is, the less dense the air, the leas drag and vice versa. Factors affecting drag/ density are temperature, altitude/ barometric pressure, humidity, efficiency of the bullet, and wind.

(1) *Temperature.* The higher the temperature, the less dense the air. (See Section III.) If the sniper zeros at 60 degrees F and he fires at 80 degrees, the air is leas dense, thereby causing an increase in muzzle velocity and higher point of impact. A 20-degree change equals a one-minute elevation change in the strike of the bullet.

(2) *Altitude/barometric pressure.* Since the air pressure is less at higher altitudes, the air is less dense. Thus, the bullet is more efficient and impacts higher due to less drag. (Table 3-1 shows the approximate

effect of change of the point of impact from sea level to 10,000 feet if the rifle is zeroed at sea level.) Impact will be the point of aim at sea level. For example, a rifle zeroed at sea level and fired at a range of 700 meters at an altitude of 5,000 feet will hit 1.6 minutes high.

RANGE (METERS)	2,500 FEET *(ASL)	5,000 FEET (ASL)	10,000 FEET (ASL)
100	.05	.08	.13
200	.1	.2	.34
300	.2	.4	.6
400	.4	.5	.9
500	.5	.9	1.4
600	.6	1.0	1.8
700	1.0	1.6	2.4
800	1.3	1.9	3.3
900	1.6	2.8	4.8
1,000	1.8	3.7	6.0
*ABOVE SEA LEVEL			

Table 3-1. Point of impact rises as altitude increases (data are in MOA).

(3) *Humidity.* Humidity varies along with the altitude and temperature. Figure 3-19 considers the changes in altitudes. Problems can occur if extreme humidity changes exist in the area of operations. That is, when humidity goes up, impact goes down; when humidity goes down, impact goes up. Since impact is affected by humidity, a 20 percent change in humidity equals about one minute as a rule of thumb. Keeping a good sniper data book during training and acquiring experience are the best teachers.

(4) *Efficiency of the bullet.* This is called a *bullet's ballistic coefficient.* The imaginary perfect bullet is rated as being 1.00. Match bullets range

from .500 to about .600. The 7.62-mm special ball (M118) is rated at .530 (Table 3-2).

(5) *Wind.* Wind is discussed in Section III.

RANGE (METERS)	(A)	(B)	(C)	(D)
100	2,407	.7	NA	.1
200	2,233	3.0	1.5	.2
300	2,066	7.3	3.0	.4
400	1,904	14.0	3.5	.5
500	1,750	24.0	4.0	.7
600	1,603	37.6	4.5	.9
700	1,466	56.2	5.0	1.0
800	1,339	80.6	5.0	1.3
900	1,222	112.5	6.0	1.5
1,000	1,118	153.5	7.0	1.8

(A) RETAINED VELOCITY (FEET PER SECOND).
(B) MIDRANGE TRAJECTORY (INCHES).
(C) BULLET DROP IN 100-METER INCREMENTS (MINUTES).
(D) TIME OF FLIGHT (SECONDS).

Table 3-2. Muzzle velocity data for 7.62-mm special ball (M118).

3-11. ANGLE FIRING

Most practice firing conducted by the sniper team involves the use of military range facilities, which are relatively flat. However, as a sniper being deployed to other regions of the world, the chance exists for operating in a mountainous or urban environment. This requires target engagements at higher and lower elevations. Unless the sniper takes corrective action, bullet impact will be above the point of aim. How high the bullet hits is determined by the range and angle to the target (Table 3-3). The amount of elevation change applied to the telescope of the rifle for angle firing is known as *slope dope.*

RANGE (METERS)	SLANT DEGREES											
	5	10	15	20	25	30	35	40	45	50	55	60
100	.01	.04	.09	.16	.25	.36	.49	.63	.79	.97	1.2	1.4
200	.03	.09	.2	.34	.53	.76	1.	1.3	1.7	2.	2.4	2.9
300	.03	.1	.3	.5	.9	1.2	1.6	2.1	2.7	3.2	3.9	4.5
400	.05	.19	.43	.76	1.2	1.7	2.3	2.9	3.7	4.5	5.4	6.3
500	.06	.26	.57	1.	1.6	2.3	3.	3.9	4.9	6.	7.2	8.4
600	.08	.31	.73	1.3	2.	2.9	3.9	5.	6.3	7.7	9.2	10.7
700	.1	.4	.9	1.6	2.5	3.6	4.9	6.3	7.9	9.6	11.5	13.4
800	.13	.5	1.	2.	3.	4.4	5.9	7.7	9.6	11.7	14.	16.4
900	.15	.6	1.3	2.4	3.7	5.3	7.2	9.3	11.6	14.1	16.9	19.8
1,000	.2	.7	1.6	2.8	4.5	6.4	8.6	11.	13.9	16.9	20.2	23.7

*RANGE GIVEN IS SLANT RANGE (METERS), NOT MAP DISTANCE.

Table 3-3. Bullet rise at given angle and range in minutes.

Section III
EFFECTS OF WEATHER

For the highly trained sniper, the effects of weather are the main causes of error in the strike of the bullet. Wind, mirage, light, temperature, and humidity affect the bullet, the sniper, or both. Some effects are minor; however, sniping is often done in extremes of weather and all effects must be considered.

3-12. WIND CLASSIFICATION

Wind poses the biggest problem for the sniper. The effect that wind has on the bullet increases with range. This is due mainly to the slowing of the bullet's velocity combined with a longer flight time. This allows the wind to have a greater effect on the round as distances increase. The result is a loss of stability.

a. Wind also has a considerable effect on the sniper. The stronger the wind, the more difficult it is for him to hold the rifle steady. This can be partly offset by training, conditioning and the use of supported positions.

b. Since the sniper must know how much effect the wind will have on the bullet, he must be able to classify the wind. The best method is to use the clock system (Figure 3-19). With the sniper at the center of the clock and the target at 12 o'clock, the wind is assigned three values: full, half, and no value. Full value means that the force of the wind will have a full effect on the flight of the bullet. These winds come from 3 and 9 o'clock. Half value means that a wind at the same speed, but from 1,2,4,5,7,8, 10, and 11 o'clock, will move the bullet only half as much as a full-value wind. No value means that a wind from 6 or 12 o'clock will have little or no effect on the flight of the bullet.

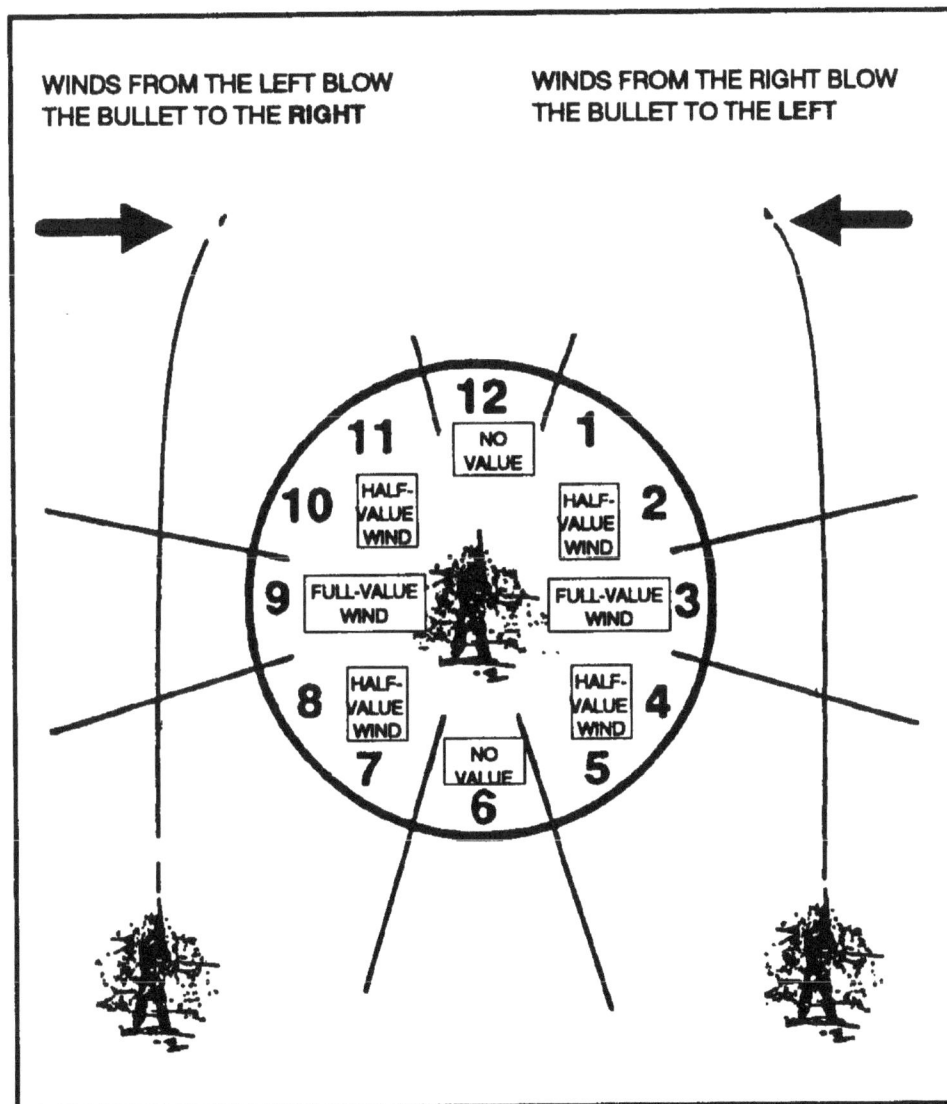

WINDS FROM THE LEFT BLOW THE BULLET TO THE **RIGHT**

WINDS FROM THE RIGHT BLOW THE BULLET TO THE **LEFT**

12
11   NO VALUE   1
10   HALF-VALUE WIND   HALF-VALUE WIND   2
9   FULL-VALUE WIND   FULL-VALUE WIND   3
8   HALF-VALUE WIND   HALF-VALUE WIND   4
7   NO VALUE   5
6

Figure 3-19. Clock system.

3-13. WIND VELOCITY

Before adjusting the sight to compensate for wind, the sniper must determine wind direction and velocity. He may use certain indicators to accomplish this. These are range flags, smoke, trees, grass, rain, and the sense of feel. However, the preferred method of determining wind direction and velocity is reading mirage (see paragraph d below). In most cases, wind direction can be determined simply by observing the indicators.

a. A common method of estimating the velocity of the wind during training is to watch the range flag (Figure 3-20). The sniper determines the angle between the flag and pole, in degrees, then divides by the constant number 4. The result gives the approximate velocity in miles per hour.

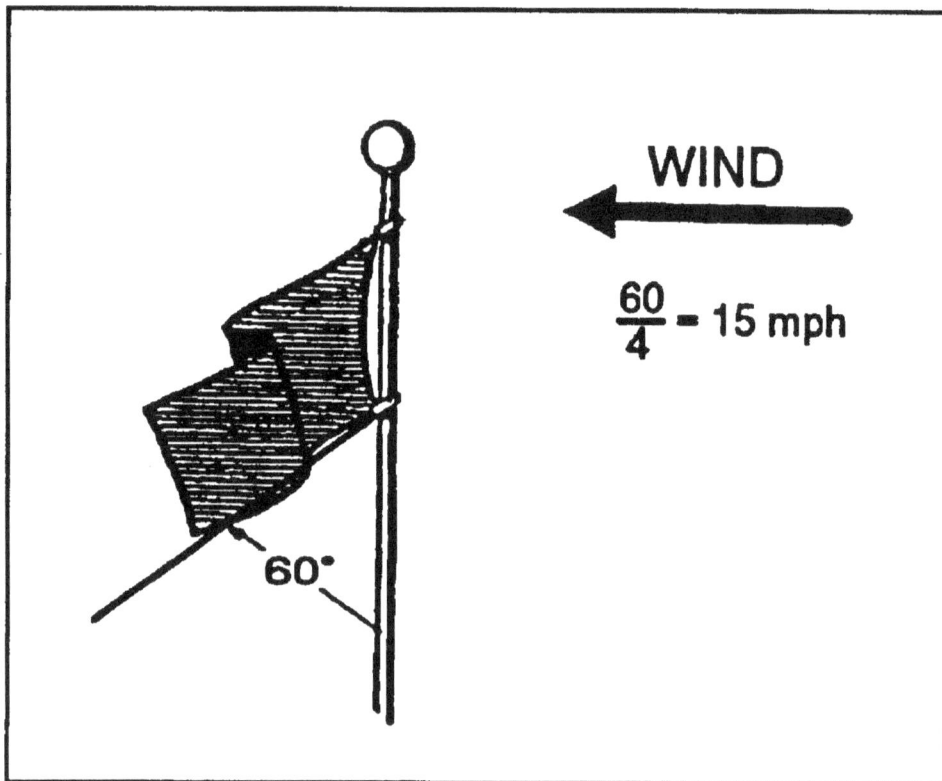

WIND

$$\frac{60}{4} = 15 \text{ mph}$$

60°

Figure 3-20. The Flag method.

b. If no flag is visible, the sniper holds a piece of paper, grass, cotton, or some other light material at shoulder level, then drops it. He then points directly at the spot where it lands and divides the angle between his body and arm by the constant number 4. This gives him the approximate wind velocity in miles per hour.

c. If these methods cannot be used, the following information is helpful in determining velocity. Winds under 3 miles per hour can barely be felt, although smoke will drift. A3- to 5-mile-per-hourwind can barely be felt on the face. With a 5- to 8-mile-per-hour wind, the leaves in the trees are in constant motion, and with a 12- to 15-mile-per-hour wind, small trees begin to sway.

d. A mirage is a reflection of the heat through layers of air at different temperatures and density as seen on a warm day (Figure 3-21). With the telescope, the sniper can see a mirage as long as there is a difference in ground and air temperatures. Proper reading of the mirage enables the sniper to estimate wind speed and direction with a high degree of accuracy. The sniper uses the M49 observation telescope to read the mirage. Since the wind nearest to midrange has the greatest effect on the bullet, he tries to determine velocity at that point. He can do this in one of two ways:

(1) He focuses on an object at midrange, then places the scope back onto the target without readjusting the focus.

(2) He can also focus on the target, then back off the focus one-quarter turn counterclockwise. This makes the target appear fuzzy, but the mirage will be clear.

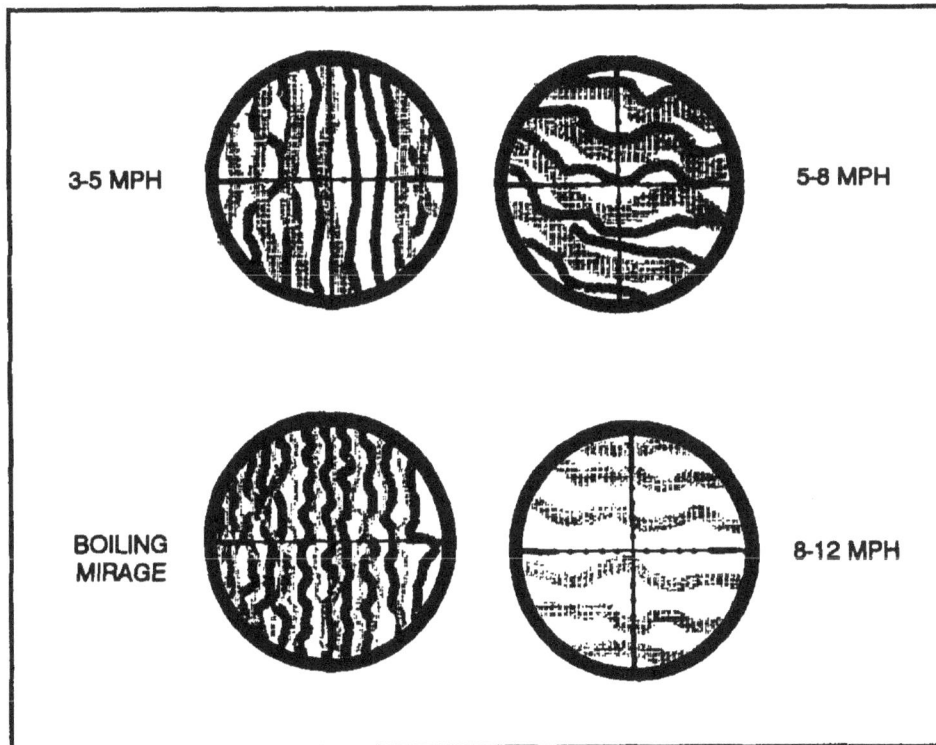

Figure 3-21. Types of mirages.

e. As observed through the telescope, the mirage appears to move with the same velocity as the wind, except when blowing straight into or away from the scope. Then, the mirage gives the appearance of moving straight upward with no lateral movement. This is called a *boiling mirage.* A boiling mirage may also be seen when the wind is constantly changing direction. For example, a full-value wind blowing from 9 o'clock to 3 o'clock suddenly changes direction. The mirage will appear to stop moving from left to right and present a boiling appearance. When this occurs, the inexperienced observer directs the sniper to fire with the "0" wind. As the sniper fires, the wind begins blowing from 3 o'clock to 9 o'clock, causing the bullet to miss the target therefore, firing in a "boil" can hamper shot placement. Unless there is a no-value wind, the sniper must wait until the boil disappears. In general, changes in the velocity of the wind, up to about 12 miles per hour, can be readily determined by observing the mirage. Beyond that speed, the movement of the mirage is too fast for detection of minor changes.

3-14. CONVERSION OF WIND VELOCITY TO MINUTES OF ANGLE
All telescopic sights have windage adjustments that are graduated in minutes of angle or fractions thereof. A minute of angle is 1/60th of a degree (Figure 3-22, page 3-34). This equals about 1 inch (1.145 inches) for every 100 meters.

EXAMPLE
1 MOA = 2 inches at 200 meters
1 MOA = 5 inches at 500 meters

a. Snipers use minutes of angle (Figure 3-22, page 3-34) to determine and adjust the elevation and windage needed on the weapon's scope. After finding the wind direction and velocity in miles per hour, the sniper must then convert it into minutes of angle, using the wind formula as a rule of thumb only. The wind formula is—

$$\frac{\text{RANGE (hundreds) divided by 100} \times \text{VELOCITY (mph)}}{\text{CONSTANT}} = \text{Minutes full-value wind}$$

The constant depends on the target's range.

```
100 to 500   "C" =15
600          "C" =14
700 to 800   "C" =13
900          "C" =12
1,000        "C" =11
```

If the target is 700 meters away and the wind velocity is 10 mph, the formula is—

$$\frac{7 \times 10}{13} = 5.38 \text{ minutes or } 5 \text{ } 1/2 \text{ minutes}$$

This determines the number of minutes for a full-value wind. For a half-value wind, the 5.38 would be divided in half.

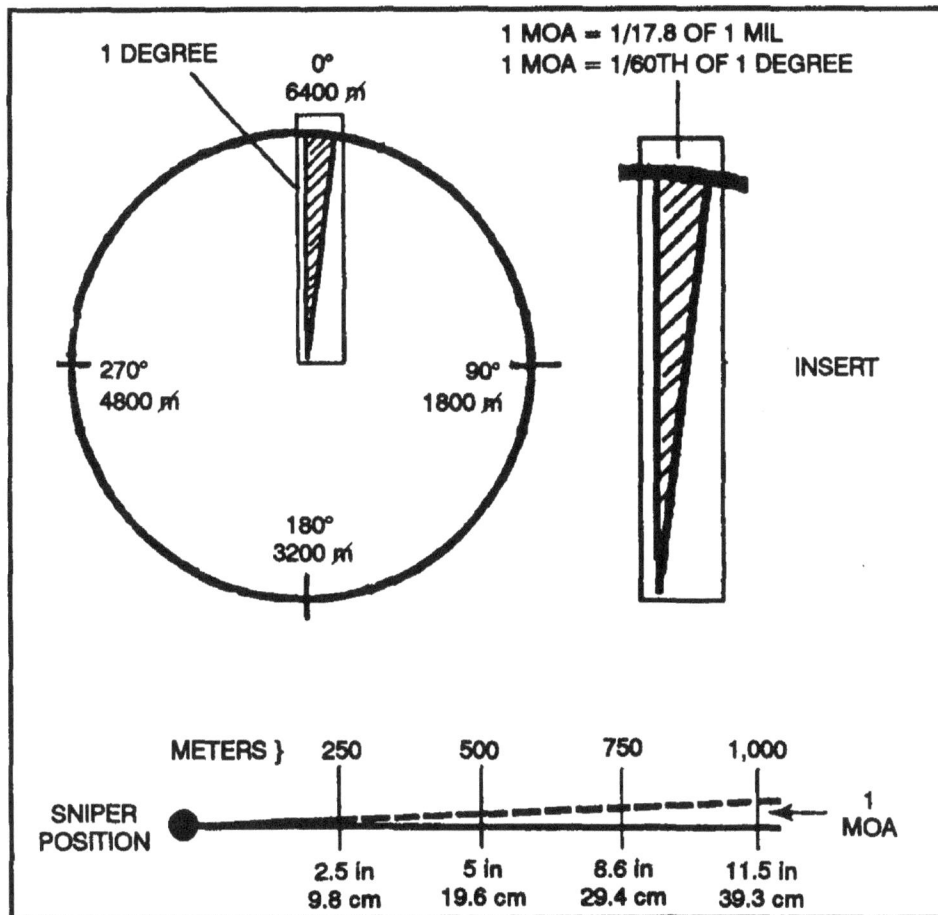

Figure 3-22. Minutes of angle.

b. The observer makes his own adjustment estimations, then compares them to the wind conversion table, which can be a valuable training tool. He must not rely on this table; if it is lost, his ability to perform the mission could be severely hampered. Until the observer gains skill in estimating wind speed and computing sight changes, he may refer to Table 3-4.

RANGE (METERS)	WIND VALUE	3 MPH MIN	IN	5 MPH MIN	IN	7 MPH MIN	IN	10 MPH MIN	IN
200	HALF	0.0	0.4	0.5	0.6	0.5	0.8	0.5	1.2
	FULL	0.5	0.8	0.5	1.2	1.0	1.7	1.0	2.4
300	HALF	0.5	0.9	0.5	1.3	0.5	1.9	1.0	2.7
	FULL	0.5	1.7	1.0	2.7	1.0	3.6	1.5	5.4
400	HALF	0.5	1.4	0.5	2.4	1.0	3.3	1.0	4.8
	FULL	0.5	2.9	1.0	4.8	1.5	6.7	2.0	9.6
500	HALF	0.5	2.3	0.5	3.8	1.0	5.3	1.5	7.5
	FULL	1.0	4.5	1.5	7.5	2.0	10.5	2.5	15.0
600	HALF	0.5	3.0	1.0	5.0	1.0	8.0	1.5	11.0
	FULL	1.0	7.0	1.5	11.0	2.5	15.0	3.5	21.0
700	HALF	0.5	4.0	'1.0	7.0	1.5	10.0	2.0	15.0
	FULL	1.0	9.0	2.0	15.0	2.5	21.0	4.0	29.0
800	HALF	0.5	6.0	1.0	10.0	1.5	13.0	2.0	19.0
	FULL	1.5	11.0	2.0	19.0	3.0	27.0	4.5	38.0
900	HALF	0.5	7.0	1.0	12.0	1.5	17.0	2.5	24.0
	FULL	3.5	15.0	2.5	24.0	3.5	34.0	5.0	49.0
1000	HALF	1.0	9.0	1.5	15.0	2.0	21.0	2.5	3.00
	FULL	1.5	18.0	2.5	30.0	4.0	42.0	5.5	60.0

RANGE (METERS)	WIND VALUE	12 MPH MIN	IN	15 MPH MIN	IN	18 MPH MIN	IN	20 MPH MIN	IN
200	HALF	0.5	1.3	1.0	1.8	1.0	2.2	1.0	2.4
	FULL	1.5	2.9	1.5	3.6	2.0	4.3	2.0	4.8
300	HALF	1.0	3.3	1.0	4.0	1.5	4.9	1.5	5.4
	FULL	2.0	6.5	2.5	8.1	3.0	9.8	3.5	10.9
400	HALF	1.5	5.8	1.5	7.2	2.0	8.6	2.0	9.6
	FULL	2.5	11.5	3.5	14.4	4.0	17.3	4.5	19.2
500	HALF	1.5	9.0	2.0	11.3	2.5	13.5	2.5	15.0
	FULL	3.5	18.0	4.0	22.6	5.0	27.0	5.5	30.0
600	HALF	1.5	13.0	2.5	16.0	3.0	19.0	3.5	22.0
	FULL	4.0	26.0	5.0	32.0	6.0	39.0	6.5	43.0
700	HALF	2.5	18.0	3.0	22.0	3.5	26.0	4.0	29.0
	FULL	4.5	35.0	6.0	44.0	7.0	53.0	7.5	59.0
800	HALF	2.5	23.0	3.5	29.0	4.0	35.0	4.5	38.0
	FULL	5.5	46.0	6.5	57.0	8.0	69.0	9.0	77.0
900	HALF	3.0	29.0	3.5	36.0	4.5	44.0	5.0	49.0
	FULL	6.0	58.0	7.5	73.0	9.0	97.0	10.0	97.0
1000	HALF	3.5	36.0	4.0	45.0	5.0	54.0	5.5	60.0
	FULL	6.5	72.0	8.0	90.0	10.0	103.0	11.5	120.0

Table 3-4. Wind conversion table.

3-15. EFFECTS OF LIGHT

Light does not affect the trajectory of the bullet; however, it does affect the way the sniper sees the target through the scope. This effect can be compared to the refraction (bending) of light through a medium, such as a prism or a fish bowl. The same effect, although not as drastic, can be observed on a day with high humidity and with sunlight from high angles. The only way the sniper can adjust for this effect is to refer to past firing recorded in the sniper data book. He can then compare different light and humidity conditions and their effect on marksmanship. Light may also affect firing on unknown distance ranges since it affects range determination capabilities.

3-16. EFFECTS OF TEMPERATURE

Temperature affects the firer, ammunition, and density of the air. When ammunition sits in direct sunlight, the bum rate of powder is increased, resulting in greater muzzle velocity and higher impact. The greatest effect is on the density of the air. As the temperature rises, the air density is lowered. Since there is leas resistance, velocity increases and once again the point of impact rises. This is in relation to the temperature at which the rifle was zeroed, If the sniper zeros at 50 degrees and he is now firing at 90 degrees, the point of impact rises considerably. How high it rises is best determined once again by past firing recorded in the sniper data book. The general role, however, is that when the rifle is zeroed, a 20-degree increase in temperature will raise the point of impact by one minute; conversely, a 20-degree decrease will drop the point of impact by one minute.

3-17. EFFECTS OF HUMIDITY

Humidity varies along with the altitude and temperature. The sniper can encounter problems if drastic humidity changes occur in his area of operation. Remember, if humidity goes up, impact goes down; if humidity goes down, impact goes up. As a rule of thumb, a 20-percent change will equal about one minute, affecting the point of impact. The sniper should keep a good sniper data book during training and refer to his own record.

Section IV
SNIPER DATA BOOK

The sniper data book contains a collection of data cards. The sniper uses the data cards to record firing results and all elements that had an effect on firing the weapon. This can vary from information about weather conditions to the attitude of the firer on that particular day. The sniper can refer to this information later to understand his weapon, the weather effects, and his shooting ability on a given day. One of the most important items of information he will record is the cold barrel zero of his weapon. A cold barrel zero refers to the first round fired from the weapon at a given range. It is critical that the sniper shoots the first round daily at different ranges. For example, Monday, 400 meters; Tuesday, 500 meters; Wednesday, 600 meters. When the barrel warms up, later shots begin to group one or two minutes higher or lower, depending on specific rifle characteristics. Information is recorded on DA Form 5785-R (Sniper's Data Card) (Figure 3-23). (A blank copy of this form is located in the back of this publication for local reproduction.)

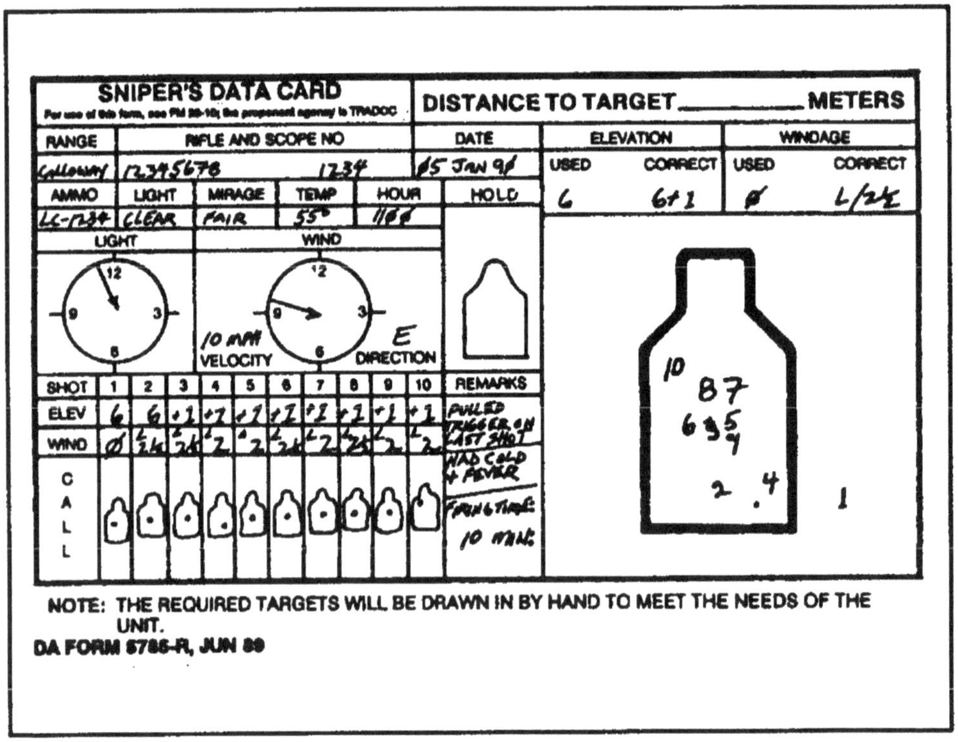

SNIPER'S DATA CARD

DISTANCE TO TARGET ________ METERS

RANGE | RIFLE AND SCOPE NO | DATE | ELEVATION (USED, CORRECT) | WINDAGE (USED, CORRECT)

AMMO | LIGHT | MIRAGE | TEMP | HOUR | HOLD

LIGHT | WIND | VELOCITY | DIRECTION

SHOT 1 2 3 4 5 6 7 8 9 10 | REMARKS

ELEV

WIND

CALL

NOTE: THE REQUIRED TARGETS WILL BE DRAWN IN BY HAND TO MEET THE NEEDS OF THE UNIT.

DA FORM 5785-R, JUN 89

Figure 3-23. Example of completed DA Form 5785-R.

3-18. ENTRIES

Three phases in writing information on the data card (Figure 3-23) are *before firing, during firing,* and *after firing.*

a. **Before Firing.** Information that is written before firing is—

(1) ***Range.*** The distance to the target.

(2) ***Rifle and scope number.*** The serial numbers of the rifle and scope.

(3) *Date.* Date of firing.

(4) Ammunition. Type and lot number of ammunition.

(5) *Light.* Amount of light (overcast, clear, and so forth).

(6) *Mirage.* Whether a mirage can be seem or not (good, bad, fair, and so forth).

(7) *Temperature.* Temperature on the range.

(8) *Hour.* Time of firing.

(9) *Light (diagram).* Draw an arrow in the direction the light is shining.

(10) *Wind.* Draw an arrow in the direction the wind is blowing, and record its average velocity and cardinal direction (N, NE, S, SW, and so forth).

b. **During Firing.** Information that is written while firing is—

(1) *Elevation.* Elevation setting used and any correction needed. For example: The target distance is 600 meters; the sniper sets the elevation dial to 6. The sniper fires and the round hits the target 6 inches low of center. He then adds one minute (one click) of elevation (+1).

(2) *Windage.* Windage setting used and any correction needed. For example The sniper fires at a 600-meter target with windage setting on 0; the round impacts 15 inches right of center. He will then add 2 1/2 minutes left to the windage dial (L/2 1/2).

(3) *Shot.* The column of information about a particular shot. For example: Column 1 is for the first round; column 10 is for the tenth round.

(4) *Elevation.* Elevation used (6 +1, 6,6 –1, and so on).

(5) *Wind.* Windage used (L/ 2 1/ 2, O, R/ l/ 2, and so on).

(6) *Call.* Where the aiming point was when the weapon fired.

(7) *Large silhouette.* Used to record the exact impact of the round on the target. This is recorded by writing the shot's number on the large silhouette in the same place it hit the target.

c. **After Firing.** After firing, the sniper records any comments about firing in the remarks section. This can be comments about the weapon, firing conditions (time allowed for fire), or his condition (nervous, felt bad, felt good, and so forth).

3-19. ANALYSIS

When the sniper leaves the firing line, he compares weather conditions to the information needed to hit the point of aim/ point of impact. Since he fires in all types of weather conditions, he must be aware of temperature, light, mirage, and wind. The sniper must consider other major points or tasks to complete

a. Compare sight settings with previous firing sessions. If the sniper always has to fine-tune for windage or elevation, there is a chance he needs a sight change (slip a scale).

b. Compare ammunition by lot number for best rifle and ammunition combination.

c. Compare all groups fired under each condition. Check the low and high shots as well as those to the left and the right of the main group—the less dispersion, the better. If groups are tight, they are easily moved to the center of the target; if loose, there is a problem. Check the scope focus and make sure the rifle is cleaned correctly. Remarks in the sniper data book will also help.

d. Make corrections. Record corrections in the sniper data book, such as position and sight adjustment information, to ensure retention.

e. Analyze a group on a target. This is important for marksmanship training. The firer may not notice errors during firing, but errors become apparent when analyzing a group. This can only be done if the sniper data book has been used correctly. A checklist that will aid in shot group/performance analysis follows:

(1) Group tends to be low and right.
- Left hand not positioned properly.
- Right elbow slipping.
- Improper trigger control.

(2) Group scattered about the target.
- Incorrect eye relief or sight picture.
- Concentration on the target (iron sights).
- Stock weld changed.
- Unstable firing position.

(3) Good group but with several erratic shots.
- *Flinching.* Shots may be anywhere.
- *Bucking.* Shots from 7 to 10 o'clock.
- *Jerking.* Shots may be anywhere.

(4) Group strung up and down through the target.
- Breathing while firing.
- Improper vertical alignment of cross hairs.
- Stock weld changed.

(5) Compact group out of the target.
- Incorrect zero.
- Failure to compensate for wind.
- Bad natural point of aim.
- Scope shadow.

(6) Group center of the target out the bottom.
- Scope shadow.
- Position of the rifle changed in the shoulder.

(7) Horizontal group across the target.
- Scope shadow.
- Canted weapon.
- Bad natural point of aim.

Section V
HOLDOFF

Holdoff is shifting the point of aim to achieve a desired point of impact. Certain situations, such as multiple targets at varying ranges and rapidly changing winds, do not allow proper windage and elevation adjustments. Therefore, familiarization and practice of elevation and windage holdoff techniques prepare the sniper to meet these situations.

3-20. ELEVATION

This technique is used only when the sniper does not have time to change his sight setting. The sniper rarely achieves pinpoint accuracy when holding off, since a minor error in range determination or a lack of a precise aiming point might cause the bullet to miss the desired point. He uses holdoff with the sniperscope only if several targets appear at various ranges, and time does not permit adjusting the scope for each target.

a. The sniper uses holdoff to hit a target at ranges other than the range for which the rifle is presently adjusted. When the sniper aims directly at a target at ranges greater than the set range, his bullet will hit below the point of aim. At lesser ranges, his bullet will hit higher than the point of aim. If the sniper understands this and knows about trajectory and bullet drop, he will be able to hit the target at ranges other than that for which the rifle was adjusted. For example, the sniper adjusts the rifle for a target located 500 meters downrange and another target appears at a range of 600 meters. The holdoff would be 25 inches, that is, the sniper should hold off 25 inches above the center of visible mass in order to hit the center of mass of that particular target (Figure 3-24). If another target were to appear at 400 meters, the sniper would aim 14 inches below the ureter of visible mass in order to hit the center of mass (Figure 3-25).

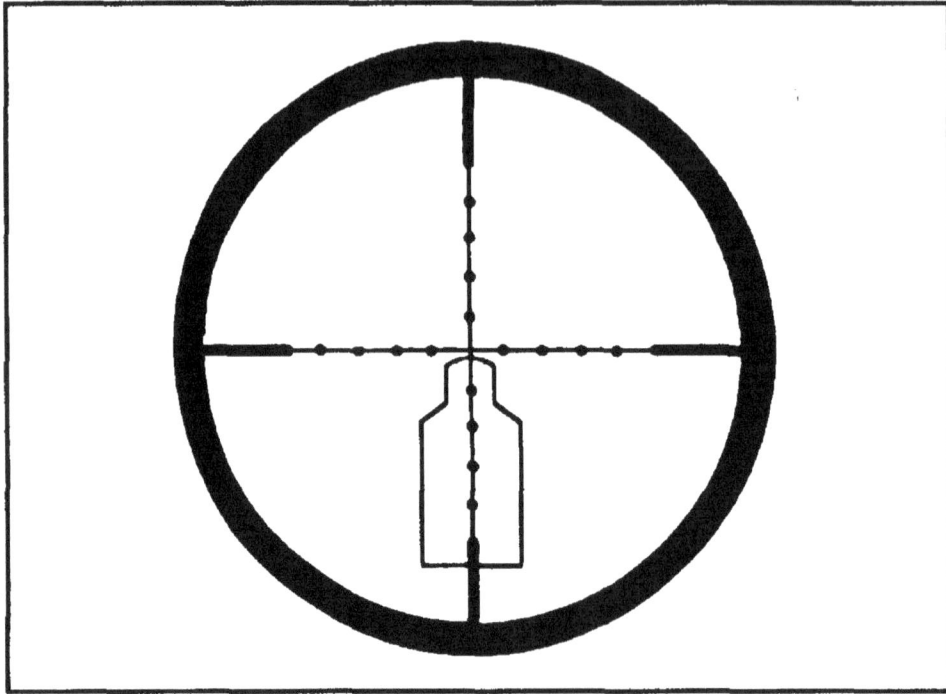

Figure 3-24. Elevation.

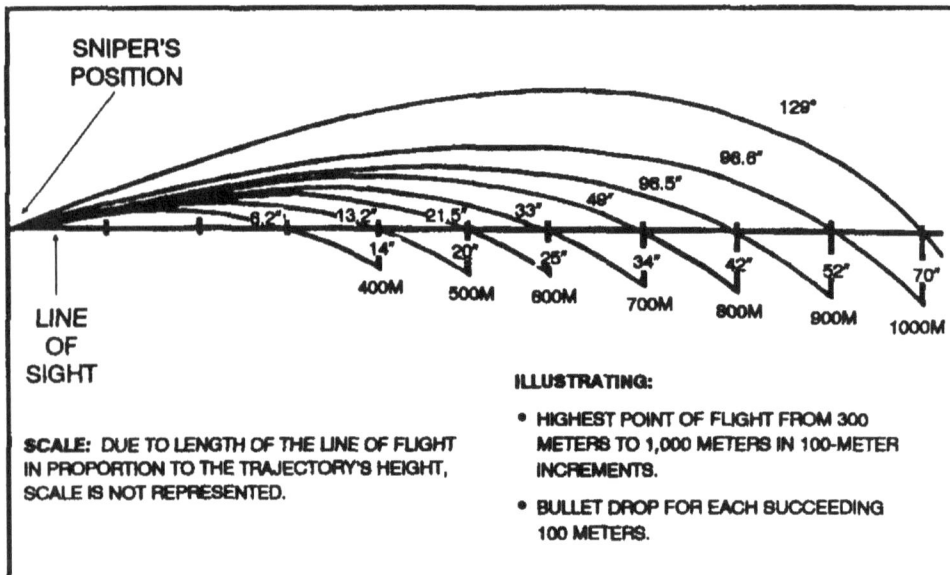

Figure 3-25. Trajectory chart.

b. The vertical mil dots on the M3A scope's reticle can be used as aiming points when using elevation holdoffs. For example, if the sniper has to engage a target at 500 meters and the scope is set at 400 meters, he would place the first mil dot 5 inches below the vertical line on the target's center mass. This gives the sniper a 15-inch holdoff at 500 meters.

3-21. WINDAGE

The sniper can use holdoff in three ways to compensate for the effect of wind.

a. When using the M3A scope, the sniper uses the horizontal mil dots on the reticle to hold off for wind. For example, if the sniper has a target at 500 meters that requires a 10-inch holdoff, he would place the target's center mass halfway between the cross hair and the first mil dot (1/2 mil) (Figure 3-26).

b. When holding off, the sniper aims into the wind. If the wind is moving from the right to left, his point of aim is to the right. If the wind is moving from left to right, his point of aim is to the left.

c. Constant practice in wind estimation can bring about proficiency in making sight adjustments or learning to apply holdoff correctly. If the sniper misses the target and the point of impact of the round is observed, he notes the lateral distance of his error and refires, holding off that distance in the opposite direction.

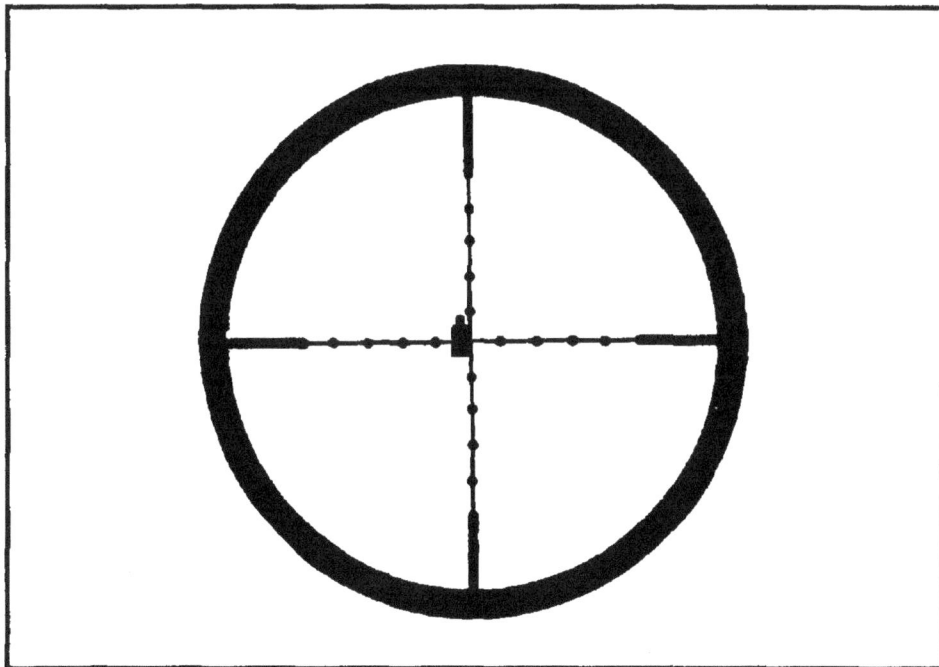

Figure 3-26. Holdoff for 7.62-mm special ball (M118).

Section VI
ENGAGEMENT OF MOVING TARGETS

Engaging moving targets not only requires the sniper to determine the target distance and wind effects on the round, but he must also consider the lateral and speed angle of the target, the round's time of flight, and the placement of a proper lead to compensate for both. These added variables increase the chance of a miss. Therefore, the sniper should engage moving targets when it is the only option.

3-22. TECHNIQUES
To engage moving targets, the sniper employs the following techniques:
- Leading.
- Tracking.
- Trapping or ambushing.
- Tracking and holding.
- Firing a snap shot.

a. **Leading.** Engaging moving targets requires the sniper to place the cross hairs ahead of the target's movement. The distance the cross hairs are placed in front of the target's movement is called a *lead.* There are four factors in determining leads:

(1) *Speed of the target. As* a target moves faster, it will move a greater distance during the bullet's flight. Therefore, the lead increases as the target's speed increases.

(2) *Angle of movement.* A target moving perpendicular to the bullet's flight path moves a greater lateral distance than a target moving at an angle away from or toward the bullet's path. Therefore, a target moving at a 45-degree angle covers less ground than a target moving at a 90-degree angle.

(3) *Range to the target.* The farther away a target is, the longer it takes for the bullet to reach it. Therefore, the lead must be increased as the distance to the target increases.

(4) *Wind effects.* The sniper must consider how the wind will affect the trajectory of the round. A wind blowing against the target's direction of movement requires less of a lead than a wind blowing in the same direction as the target's movement.

b. Tracking. hacking requires the sniper to establish an aiming point ahead of the target's movement and to maintain it as the weapon is fired. This requires the weapon and body position to be moved while following the target and firing.

c. **Trapping or Ambushing.** Trapping or ambushing is the sniper's preferred method of engaging moving targets. The sniper must establish an aiming point ahead of the target and pull the trigger when the target reaches it. This method allows the sniper's weapon and body position to remain motionless. With practice, a sniper can determine exact leads and aiming points using the horizontal stadia lines in the mil dots in the M3A.

d. **Tracking and Holding.** The sniper uses this technique to engage an erratically moving target. That is, while the target is moving, the sniper keeps his cross hairs centered as much as possible and adjusts his position with the target. When the target stops, the sniper quickly perfects his hold and fires. This technique requires concentration and discipline to keep from firing before the target comes to a complete halt.

e. **Firing a Snap Shot.** A sniper may often attempt to engage a target that only presents itself briefly, then resumes cover. Once he establishes a pattern, he can aim in the vicinity of the target's expected appearance and fire a snap shot at the moment of exposure.

3-23. COMMON ERRORS

When engaging moving targets, the sniper makes common errors because he is under greater stress than with a stationary target. There are more considerations, such as retaining a steady position and the correct aiming point, how fast the target is moving, and how far away it is. The more practice a sniper has shooting moving targets, the better he will become. Some common mistakes are as follows:

a. The sniper has a tendency to watch his target instead of his aiming point. He must force himself to watch his lead point.

b. The sniper may jerk or flinch at the moment his weapon fires because he thinks he must fire NOW. This can be overcome through practice on a live-fire range.

c. The sniper may hurry and thus forget to apply wind as needed. Windage must be calculated for moving targets just as for stationary targets. Failure to do this when squiring a lead will result in a miss.

3-24. CALCULATION OF LEADS

Once the required lead has been determined, the sniper should use the mil scale in the scope for precise holdoff. The mil scale can be mentally sectioned into 1/4-mil increments for leads. The chosen point on the mil scale becomes the sniper's point of concentration just as the cross hairs are for stationary targets. The sniper concentrates on the lead point and fires the weapon when the target is at this point. The following formulas are used to determine moving target leads:

TIME OF FLIGHT X TARGET SPEED = LEAD.

Time of flight= flight time of the round in seconds.

Target speed = speed the target is moving in fps.

Lead = distance aiming point must be placed ahead of movement in feet.

Average speed of a man during—

Slow patrol = 1 fps/ 0.8 mph

Fast patrol = 2 fps/ 1.3 mph

Slow walk = 4 fps/ 2.5 mph

Fast walk = 6 fps/ 3.7 mph

To convert leads in feet to meters:

LEAD IN FEET X 0,3048 = METERS

To convert leads in meters to mils:

$$\frac{\text{LEAD IN METERS} \times 1,000}{\text{RANGE TO TARGET}} = \text{MIL LEAD}$$

Section VII
NUCLEAR, BIOLOGICAL CHEMICAL

Performance of long-range precision fire is difficult at best. Enemy NBC warfare creates new problems for the sniper. Not only must the sniper properly execute the fundamentals of marksmanship and contend with the forces of nature, he must overcome obstacles presented by protective equipment. Testing conducted by the US Army Sniper School, Fort Benning, GA during 1989 to 1990 uncovered several problem areas. Evaluation of this testing discovered ways to help the sniper overcome these problems while firing in an NBC environment.

3-25. PROTECTIVE MASK
The greatest problem while firing the M24 with the M17-series protective mask was that of recoil breaking the seal of the mask. Also, due to filter elements and hard eye lenses, the sniper could not gain and maintain proper stock weld and eye relief. Additionally, the observer could

not gain the required eye relief for observation through his M49 observation telescope. However, testing of the M25-series protective mask provided the following results:

a. Because of its separate filtering canister, the stock weld was gained and maintained with minimal effort.

b. Its flexible face shield allowed for excellent observation. This also allowed the sniper and observer to achieve proper eye relief, which was needed for observation with their respective telescopes.

3-26. MISSION-ORIENTED PROTECTION POSTURE

Firing while in MOPP has a significant effect on the ability to deliver precision fire. The following problems and solutions have been identified

a. **Eye Relief.** Special emphasis must be made in maintaining proper eye relief and the absence of scope shadow. Maintaining consistent stock weld is a must.

b. **Trigger Control.** Problems encountered with trigger control consist of the sense of touch and stock drag.

(1) *Sense of touch.* When gloves are worn, the sniper cannot determine the amount of pressure he is applying to the trigger. This is of particular importance if the sniper has the trigger adjusted for a light pull. 'Raining with a glove will be beneficial; however, the trigger should be adjusted to allow the sniper to feel the trigger without accidental discharge.

(2) *Stock drag.* While training, the sniper should have his observer watch his trigger finger to ensure that the finger and glove are not touching any part of the rifle but the trigger. The glove or finger resting on the trigger guard moves the rifle as the trigger is pulled to the rear. The sniper must wear a well-fitted glove.

c. **Vertical Sight Picture.** The sniper naturally cants the rifle into the cheek of the face while firing with a protective mask.

d. **Sniper/Observer Communications.** The absence of a voice emitter on the M2S-series protective mask creates an obstacle in relaying information. The team either speaks louder or uses written messages. A system of foot taps, finger taps, or hand signals may be devised. Communication is a must; training should include the development and practice of communications at different MOPP levels.

CHAPTER 4

FIELD TECHNIQUES

The primary mission of the sniper team is to eliminate selected enemy targets with long-range precision fire. How well the sniper accomplishes his mission depends on knowledge, understanding and application of various field techniques that allow him to move, hide, observe, and detect targets. This chapter discusses the field techniques and skills that the sniper must learn before employment in support of combat operations. The sniper's application of these skills will affect his survival on the battlefield.

Section I
CAMOUFLAGE

Camouflage is one of the basic weapons of war. It can mean the difference between a successful or unsuccessful mission. To the sniper team, it can mean the difference between life and death. Camouflage measures are important since the team cannot afford to be detected at any time while moving alone, as part of another element, or while operating from a firing position. Marksmanship training teaches the sniper to hit a target, and a knowledge of camouflage teaches him how to avoid becoming a target. Paying attention to camouflage fundamentals is a mark of a well-trained sniper. (See FM 5-20 for more details.)

4-1. TARGET INDICATORS
To become proficient in camouflage, the sniper team must first understand target indicators. Target indicators are anything a soldier does or fails to do that could result in detection. A sniper team must know and understand target indication not only to move undetected, but also to detect enemy movement. Target indicators are sound, movement, improper camouflage, disturbance of wildlife, and odors.

a. Sound.
- Most noticeable during hours of darkness.
- Caused by movement, equipment rattling, or talking.
- Small noises may be dismissed as natural, but talking will not.

b. Movement.
- Most noticeable during hours of daylight.
- The human eye is attracted to movement.
- Quick or jerky movement will be detected faster than slow movement.

c. Improper camouflage.
- Shine.
- Outline.
- Contrast with the background.

d. Disturbance of wildlife.
- Birds suddenly flying away.
- Sudden stop of animal noises.
- Animals being frightened.

e. Odors.
- Cooking.
- Smoking.
- Soap and lotions.
- Insect repellents.

4-2. BASIC METHODS

The sniper team can use three basic methods of camouflage. It may use one of these methods or a combination of all three to accomplish its objective. The three basic methods a sniper team can use are hiding, blending, and deceiving.

a. **Hiding.** Hiding is used to conceal the body from observation by lying behind an object or thick vegetation.

b. **Blending.** Blending is used to match personal camouflage with the surrounding area to a point where the sniper cannot be seen.

c. **Deceiving.** Deceiving is used to fool the enemy into false conclusions about the location of the sniper team.

4-3. TYPES OF CAMOUFLAGE

The two types of camouflage that the sniper team can use are *natural* and *artificial*.

a. **Natural.** Natural camouflage is vegetation or materials that are native to the given area. The sniper augments his appearance by using natural camouflage.

b. **Artificial.** Artificial camouflage is any material or substance that is produced for the purpose of coloring or covering something in order to conceal it. Camouflage sticks or face paints are used to cover all exposed areas of skin such as face, hands, and the back of the neck. The parts of the face that form shadows should be lightened, and the parts that shine should be darkened. The three types of camouflage patterns the sniper team uses are striping, blotching, and combination.

(1) *Striping.* Used when in heavily wooded areas and when leafy vegetation is scarce.

(2) *Blotching.* Used when an area is thick with leafy vegetation.

(3) *Combination.* Used when moving through changing terrain. It is normally the best all-round pattern.

4-4. GHILLIE SUIT

The ghillie suit is a specially made camouflage uniform that is covered with irregular patterns of garnish or netting (Figure 4-1).

a. Ghillie suits can be made from BDUs or one-piece aviator-type uniforms. Turning the uniform inside out places the pockets inside the suit. This protects items in the pockets from damage caused by crawling on the ground. The front of the ghillie suit should be covered with canvas or some type of heavy cloth to reinforce it. The knees and elbows should be covered with two layers of canvas, and the seam of the crotch should be reinforced with heavy nylon thread since these areas are prone to wear out quicker.

b. The garnish or netting should cover the shoulders and reach down to the elbows on the sleeves. The garnish applied to the back of the suit should be long enough to cover the sides of the sniper when he is in the prone position. A bush hat is also covered with garnish or netting. The garnish should belong enough to breakup the outline of the sniper's neck, but it should not be so long in front to obscure his vision or hinder movement.

e. A veil can be made from a net or piece of cloth covered with garnish or netting. It covers the weapon and sniper's head when in a firing position. The veil can be sewn into the ghillie suit or carried separately. A ghillie suit does not make one invisible and is only a camouflage base. Natural vegetation should be added to help blend with the surroundings.

CANVAS CAN BE STITCHED
WITH NYLON TWINE OR GLUED
WITH RUBBERIZED CEMENT

PLACEMENT OF NETTING
AND GARNISH

Figure 4-1. Ghillie suit.

4-5. FIELD-EXPEDIENT CAMOUFLAGE

The sniper team may have to use field-expedient camouflage if other means are not available. Instead of camouflage sticks or face paint, the team may use charcoal, walnut stain, mud, or whatever works. The team will not use oil or grease due to the strong odor. Natural vegetation can be attached to the body by boot bands or rubber bands or by cutting holes in the uniform.

a. The sniper team also camouflages its equipment. However, the camouflage must not interfere with or hinder the operation of the equipment.

(1) *Rifles.* The sniper weapon system and the M16/ M203 should also. be camouflaged to break up their outlines. The sniper weapon system can be carried in a "drag bag" (Figure 4-2), which is a rifle case made of canvas and covered with garnish similar to the ghillie suit.

(2) *Optics.* Optics used by the sniper team must also be camouflaged to breakup the outline and to reduce the possibility of light reflecting off the lenses. Lenses can be covered with mesh-type webbing or nylon hose material.

(3) *ALICE pack.* If the sniper uses the ALICE pack while wearing the ghillie suit, he must camouflage the pack the same as the suit.

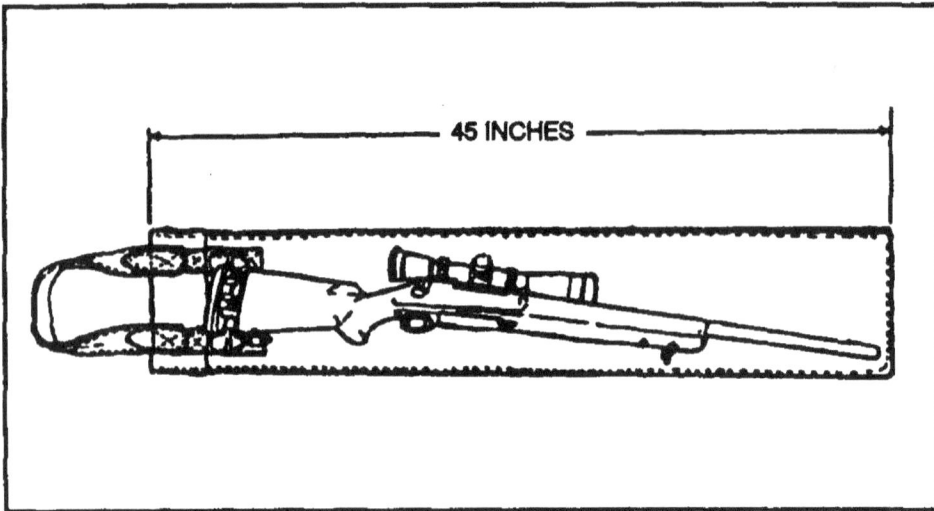

Figure 4-2. Drag bag.

b. The sniper team alters its camouflage to blend in with changes in vegetation and terrain in different geographic areas. Examples of such changes are as follows:

(1) *Snow areas.* Blending of colors is more effective than texture camouflage in snowy areas. In areas with heavy snow or in wooded areas with trees covered with snow, a full white camouflage suit should be worn. In areas with snow on the ground but not on the trees, white trousers with green and brown tops should be worn.

(2) *Desert areas.* In sandy desert areas that have little vegetation, the blending of tan and brown colors is important. In these areas, the sniper team must make full use of the terrain and the vegetation that is available to remain unnoticed.

(3) *Jungle areas.* In jungle areas, textured camouflage, contrasting colors, and natural vegetation must be used.

(4) *Urban areas.* In urban areas, the sniper team's camouflage should be a blended color (shades of gray usually work best). Texutred camouflage is not as important in these environments.

c. The sniper team must be camouflage conscious from the time it departs on a mission until it returns. It must constantly use the terrain, vegetation, and shadows to remain undetected. At no other time during the mission will the sniper team have a greater tendency to be careless than during its return to a friendly area. Fatigue and undue haste may override caution and planning. Therefore, the team needs to pay close attention to its camouflage discipline on return from missions.

4-6. COVER AND CONCEALMENT

The proper understanding and application of the principles of cover and concealment used with the proper application of camouflage protects the sniper team from enemy observation.

a. Cover is natural or artificial protection from the fire of enemy weapons. Natural cover (ravines, hollows, reverse slopes) and artificial cover (fighting positions, trenches, walls) protect the sniper team from flat trajectory fires and partly protect it from high-angle fires and the effects of nuclear explosions. Even the smallest depression or fold in the ground may provide some cover when the team needs it most. A 6-inch depression, properly used, may provide enough cover to save the sniper team under fire. Snipers must always look for and take advantage of all the cover that the terrain provides. By combining this habit with proper movement techniques, the team can protect itself from enemy fire. To get protection from enemy fire when moving, the team uses routes that put cover between itself and the enemy.

b. Concealment is natural or artificial protection from enemy observation. The surroundings may provide natural concealment that needs no change before use (bushes, grass, and shadows). The sniper team creates artificial concealment from materials such as burlap and camouflage nets, or it can move natural materials (bushes, leaves, and grass) from their original location. The sniper team must consider the effects of the change of seasons on the concealment provided by both natural and artificial materials. 'he principles of concealment include the following

(1) *Avoid unnecessary movement.* Remain still—movement attracts attention. The position of the sniper team is concealed when the team remains still, but the sniper's position is easily detected when the team moves. Movement against a stationary background makes the team stand out clearly. When the team must change positions, it moves carefully over a concealed route to a new position, preferably during limited visibility. Snipers move inches at a time, slowly and cautiously, always scanning ahead for the next position.

(2) *Use all available concealment.* Available concealment includes the following

(a) *Background.* Background is important the sniper team must blend with it to prevent detection. The trees, bushes, grass, earth, and man-made structures that form the background vary in color and appearance. This makes it possible for the team to blend with them. The team selects trees or bushes to blend with the uniform and to absorb the figure outline. Snipers must always assume they are under observation.

(b) *Shadows.* The sniper team in the open stands out clearly, but the sniper team in the shadows is difficult to see. Shadows exist under most conditions, day and night. A sniper team should never fire from the edge of a wood line; it should fire from a position inside the wood line (in the shade or shadows provided by the tree tops).

(3) *Stay low to observe.* A low silhouette makes it difficult for the enemy to see a sniper team. Therefore, the team observes from a crouch, a squat, or a prone position.

(4) *Avoid shiny reflections.* Reflection of light on a shiny surface instantly attracts attention and can be seen from great distances. The sniper uncovers his rifle scope only when indexing and aiming at a target. He uses optics cautiously in bright sunshine because of the reflections they cause.

(5) *Avoid skylining.* Figures on the skyline can be seen from a great distance, even at night, because a dark outline stands out against the lighter sky. The silhouette formed by the body makes a good target.

(6) *Alter familiar outlines.* Military equipment and the human body are familiar outlines to the enemy. The sniper team alters or disguises these revealing shapes by using the ghillie suit or outer smock that is covered with irregular patterns of garnish. The team must alter its outline from the head to the soles of the boots.

(7) *Observe noise discipline.* Noise, such as talking, can be picked up by enemy patrols or observation posts. The sniper team silences gear before a mission so that it makes no sound when the team walks or runs.

Section II
MOVEMENT

A sniper team's mission and method of employment differ in many ways from those of the infantry squad. One of the most noticeable differences is the movement technique used by the sniper team. Movement by teams must not be detected or even suspected by the enemy. Because of this, a sniper team must master individual sniper movement techniques.

4-7. RULES OF MOVEMENT
When moving, the sniper team should always remember the following rules

a. Always assume the area is under enemy observation.

b. Move slowly. A sniper counts his movement progress by feet and inches.

c. Do not cause overhead movement of trees, bushes, or tall grasses by rubbing against them.

d. Plan every movement and move in segments of the route at a time.

e. Stop, look, and listen often.

f. Move during disturbances such as gunfire, explosions, aircraft noise, wind, or anything that will distract the enemy's attention or conceal the team's movement.

4-8. INDIVIDUAL MOVEMENT TECHNIQUES

The individual movement techniques used by the sniper team are designed to allow movement without being detected. These movement techniques are sniper low crawl, medium crawl, high crawl, hand-and-knees crawl, and walking.

a. **Sniper Low Crawl.** The sniper low crawl (Figure 4-3) is used when concealment is extremely limited, when close to the enemy, or when occupying a firing position.

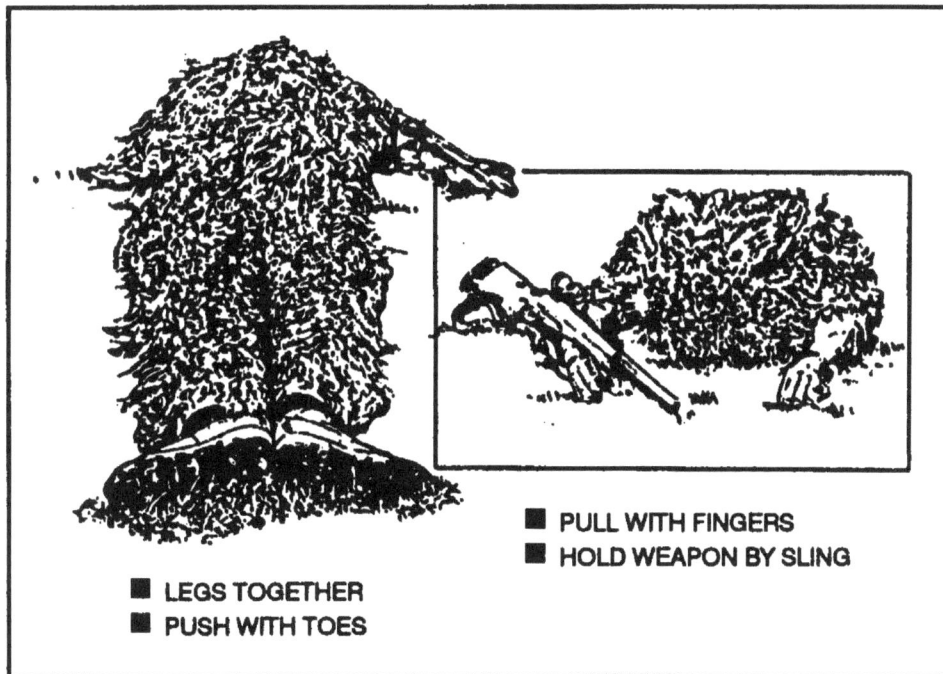

■ PULL WITH FINGERS
■ HOLD WEAPON BY SLING

■ LEGS TOGETHER
■ PUSH WITH TOES

Figure 4-3. Sniper low crawl.

b. **Medium Crawl.** The medium crawl (Figure 4-4) is used when concealment is limited and the team needs to move faster-than the sniper low crawl allows. The medium crawl is similar to the infantryman's low crawl.

c. **High Crawl.** The high crawl (Figure 4-5) is used when concealment is limited but high enough to allow the sniper to raise his body off the ground. The high crawl is similar to the infantry high crawl.

PULL WITH ARMS
HOLD WEAPON BY SLING

LIE FLAT ON GROUND
LEGS SPREAD
PUSH WITH LEGS

Figure 4-4. Medium crawl.

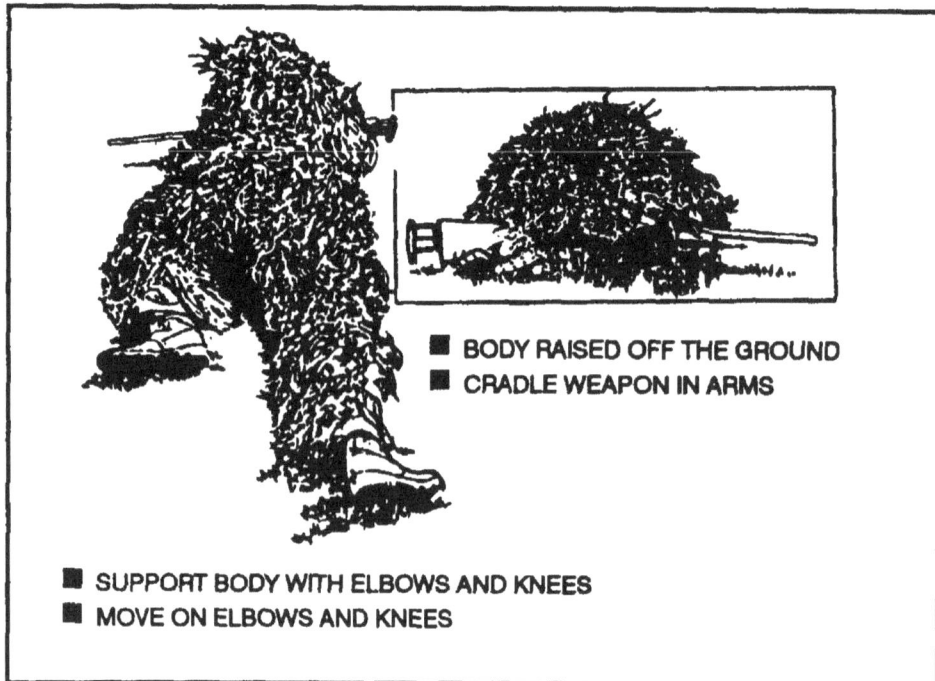

BODY RAISED OFF THE GROUND
CRADLE WEAPON IN ARMS

SUPPORT BODY WITH ELBOWS AND KNEES
MOVE ON ELBOWS AND KNEES

Figure 4-5. High crawl.

d. **Hand-and-knees Crawl.** The hand-and-knees crawl (Figure 4-6) is used when some concealment is available and the sniper team needs to move faster than the medium crawl.

- BODY SUPPORTED BY KNEES AND HAND
- WEAPON CARRIED IN OTHER HAND
- SCOPE IN ARMPIT

Figure 4-6. Hand-and-knees crawl.

e. **Walking.** Walking (Figure 4-7) is used when there is good concealment, it is not likely the enemy is close, and speed is required.

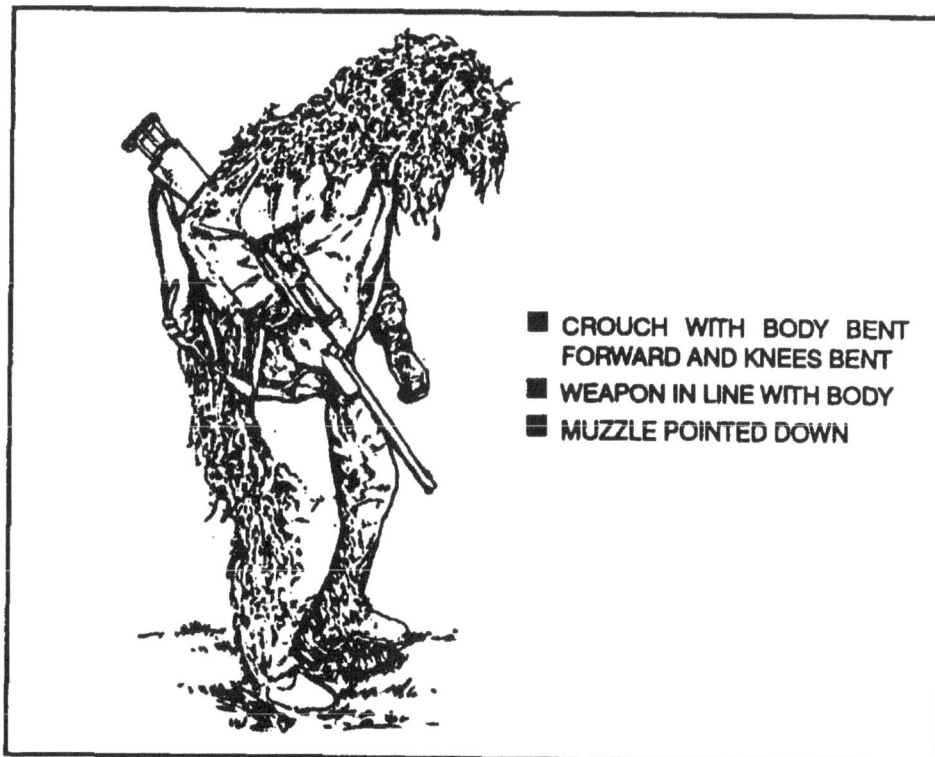

- CROUCH WITH BODY BENT FORWARD AND KNEES BENT
- WEAPON IN LINE WITH BODY
- MUZZLE POINTED DOWN

Figure 4-7. Walking.

4-9. SNIPER TEAM MOVEMENT AND NAVIGATION

Due to lack of personnel and firepower, the sniper team cannot afford detection by the enemy nor can it successfully fight the enemy in sustained engagements.

a. When possible, the sniper team should be attached to a security element (squad/platoon). The security element allows the team to reach its area of operations quicker and safer than the team operating alone. Plus, the security element provides the team a reaction force should the team be detected. Snipers use the following guidelines when attached to a security element:

(1) The security element leader is in charge of the team while it is attached to the element.

(2) The sniper team always appears as an integral part of the element.

(3) The sniper team wears the same uniform as the element members.

(4) The sniper team maintains proper intends and positions in all formations.

(5) The sniper weapon system is carried in line and close to the body, hiding its outline and barrel length.

(6) All equipment that is unique to sniper teams is concealed from view (optics, ghillie suits, and so forth).

b. Once in the area of operation, the sniper team separates from the security element and operates alone. Two examples of a sniper team separating from security elements are as follows:

(1) The security element provides security while the team prepares for operation.

(a) The team dons the ghillie suits and camouflages itself and its equipment (if mission requires).

(b) The team ensures all equipment is secure and caches any nonessential equipment (if mission requires).

(c) Once the team is prepared, it assumes a concealed position, and the security element departs the area.

(d) Once the security element has departed, the team waits in position long enough to ensure neither itself nor the security element has been compromised. Then, the team moves to its tentative position.

(2) The security element conducts a short security halt at the separation point. The sniper team halts, ensuring they have good available concealment and know each other's location. The security element then proceeds, leaving the sniper team in place. The sniper team remains in position until the security element is clear of the area. The team then organizes itself as

required by the mission and moves on to its tentative position. This type of separation also works well in MOUT situations.

c. When selecting routes, the sniper team must remember its strengths and weaknesses. The following guidelines should be used when selecting routes:

(1) Avoid known enemy positions and obstacles.

(2) Seek terrain that offers the best cover and concealment.

(3) Take advantage of difficult terrain (swamps, dense woods, and so forth).

(4) Do not use trails, roads, or footpaths.

(5) Avoid built-up or populated areas.

(6) Avoid areas of heavy enemy guerrilla activity.

d. When the sniper team moves, it must always assume its area is under enemy observation. Because of this and the size of the team with the small amount of firepower it has, the team uses only one type of formation-the sniper movement formation. Characteristics of the formation are as follows:

(1) The observer is the point man; the sniper follows.

(2) The observer's sector of security is 3 o'clock to 9 o'clock; the sniper's sector of security is 9 o'clock to 3 o'clock (overlapping).

(3) Visual contact must be maintained even when lying on the ground.

(4) An interval of no more than 20 meters is maintained.

(5) The sniper reacts to the point man's actions.

(6) The team leader designates the movement techniques and routes used.

(7) The team leader designates rally points.

e. A sniper team must never become decisively engaged with the enemy. The team must rehearse immediate action drills to the extent that they become a natural and immediate reaction should it make unexpected contact with the enemy. Examples of such actions are as follows:

(1) *Visual contact.* If the sniper team sees the enemy and the enemy does not see the team, it freezes. If the team has time, it will do the following

(a) Assume the best covered and concealed position.

(b) Remain in position until the enemy has passed.

NOTE: The team will not initiate contact.

(2) *Ambush.* In an ambush, the sniper team's objective is to break contact immediately. One example of this involves performing the following

(a) The observer delivers rapid fire on the enemy.

(b) The sniper throws smoke grenades between the observer and the enemy.

(c) The sniper delivers well-aimed shots at the most threatening targets until smoke covers the area.

(d) The observer then throws fragmentation grenades and withdraws toward the sniper, ensuring he does not mask the sniper's fire.

(e) The team moves to a location where the enemy cannot observe or place direct fire on it.

(f) If contact cannot be broken, the sniper calls for indirect fires or a security element (if attached).

(g) If team members get separated, they should return to the next-to-last designated en route rally point.

(3) *Indirect fire.* When reacting to indirect fires, the team must move out of the area as quickly as possible. This sudden movement can result in the team's exact location and direction being pinpointed. Therefore, the team must not only react to indirect fire but also take actions to conceal its movement once it is out of the impact area.

(a) The team leader moves the team out of the impact area using the quickest route by giving the direction and distance (clock method).

(b) Team members move out of the impact area the designated distance and direction.

(c) The team leader then moves the team farther away from the impact area by using the most direct concealed route. They continue the mission using an alternate route.

(d) If team members get separated, they should return to the next-to-last designated en route rally point.

(4) *Air attack.*

(a) Team members assume the best available covered and concealed positions.

(b) Between passes of aircraft, team members move to positions that offer better cover and concealment.

(c) The team does not engage the aircraft.

(d) Team members remain in positions until attacking aircraft depart.

(e) If team members get separated, they return to the next-to-last designated en route rally point.

f. To aid the sniper team in navigation, the team should memorize the route by studying maps, aerial photos, or sketches. The team notes distinctive features (hills, streams, roads) and its location in relation to the route. It plans an alternate route in case the primary route cannot be used. It plans offsets to circumvent known obstacles to movement. The team uses terrain countdown, which involves memorizing terrain features from the start point to the objective, to maintain the route. During the mission, the sniper team mentally counts each terrain feature, thus ensuring it maintains the proper route.

g. The sniper team maintains orientation at all times. As it moves, it observes the terrain carefully and mentally checks off the distinctive features noted in the planning and study of the route. Many aids are available to ensure orientation. The following are examples:

(1) The location and direction of flow of principal streams.

(2) Hills, valleys, roads, and other peculiar terrain features.

(3) Railroad tracks, power lines, and other man-made objects.

Section III
SELECTION, OCCUPATION, AND CONSTRUCTION OF SNIPER POSITIONS

Selecting the location for a position is one of the most important tasks a sniper team accomplishes during the mission planning phase of an operation. After selecting the location, the team also determines how it will move into the area to locate and occupy the final position.

4-10. SELECTION

Upon receiving a mission, the sniper team locates the target area and then determines the best location for a tentative position by using one or more of the following sources of information: topographic maps, aerial photographs, visual reconnaissance before the mission, and information gained from units operating in the area.

a. The sniper team ensures the position provides an optimum balance between the following considerations:

- Maximum fields of fire and observation of the target area.
- Concealment from enemy observation.
- Covered routes into and out of the position.
- Located no closer than 300 meters from the target area.
- A natural or man-made obstacle between the position and the target area.

b. A sniper team must remember that a position that appears to be in an ideal location may also appear that way to the enemy. Therefore, the team avoids choosing locations that are—
- On a point or crest of prominent terrain features.
- Close to isolated objects.
- At bends or ends of roads, trails, or streams.
- In populated areas, unless it is required.

c. The sniper team must use its imagination and ingenuity in choosing a good location for the given mission. The team chooses a location that not only allows the team to be effective but also must appear to the enemy to be the least likely place for a team position. The following are examples of such positions:
- Under logs in a deadfall area.
- Tunnels bored from one side of a knoll to the other.
- Swamps.
- Deep shadows.
- Inside rubble piles.

4-11. OCCUPATION

During the mission planning phase, the sniper also selects an objective rally point. From this point, the sniper team reconnoiters the tentative position to determine the exact location of its final position. The location of the ORP should provide cover and concealment from enemy fire and observation, be located as close to the selected area as possible, and have good routes into and out of the selected area.

a. From the ORP, the team moves foward to a location that allows the team to view the tentative position area (Figure 4-8 page 4-16). One member remains in this location to cover the other member who reconnoiters the area to locate a final position. Once a suitable location has been found, the covering team member moves to the position. While conducting the reconnaissance or moving to the position, the team—
- Moves slowly and deliberately, using the sniper low crawl.
- Avoids unnecessary movement of trees, bushes, and grass.
- Avoids making any noises.
- Stays in the shadows, if there are any.
- Stops, looks, and listens every few feet.

b. When the sniper team arrives at the firing position, it—
- Conducts a detailed search of the target area.
- Starts construction of the firing position, if required.

- Organizes equipment so that it is easily accessible.
- Establishes a system of observing eating resting, and latrine calls.

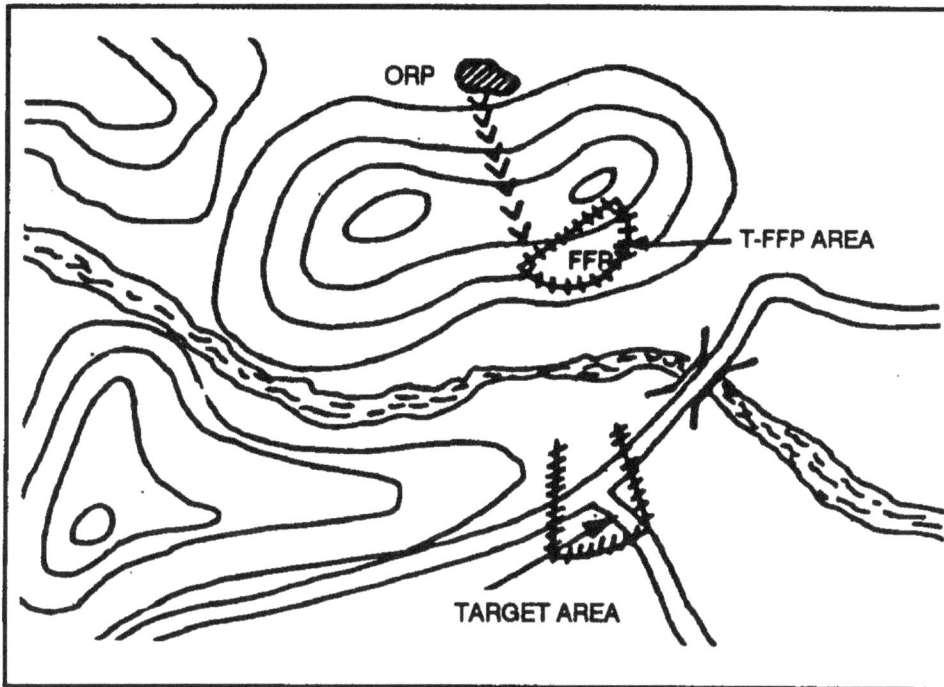

Figure 4-8. Tentative position areas.

4-12. CONSTRUCTION

A sniper mission always requires the team to occupy some type of position. These positions can range from a hasty position, which a team may use for a few hours, to a more permanent position, which the team could occupy. for a few days. The team should always plan to build its position during limited visibility.

a. **Sniper Position Considerations.** Whether a sniper team is in a position for a few minutes or a few days, the basic considerations in. choosing a type of position remain the same.

(1) *Location:*

(a) *Type of terrain and soil.* Digging and boring of tunnels can be very difficult in hard soil or in fine, loose sand. The team takes advantage of what the terrain offers (gullies, holes, hollow tree stumps, and so forth).

(b) *Enemy location and capabilities.* Enemy patrols in the area may be close enough to the position to hear any noises that may accidentally be made during any construction. The team also considers the enemy's night vision and detection capabilities.

(2) *Time:*

(a) *Amount of time to be occupied.* If the sniper team's mission requires it to be in position for a long time, the team constructs a position that provides more survivability. This allows the team to operate more effectively for a longer time.

(b) *Time required for construction.* The time required to build a position must be considered, especially during the mission planning phase.

(3) *Personnel and equipment:*

(a) *Equipment needed for construction.* The team plans for the use of any extra equipment needed for construction (bow saws, picks, axes, and so forth).

(b) *Personnel needed for construction.* Coordination is made if the position requires more personnel to build it or a security element to secure the area during construction.

b. **Construction Techniques.** Belly and semipermanent hide positions can be constructed of stone, brick, wood, or turf. Regardless of material, every effort is made to bulletproof the front of the hide position. The team can use the following techniques:

- Pack protective jackets around the loophole areas.
- Emplace an angled armor plate with a loophole cut into it behind the hide loophole.
- Sandbag the loopholes from the inside.

(1) *Pit.* Hide construction begins with the pit since it protects the sniper team. All excavated dirt is removed (placed in sandbags, taken away on a poncho, and so forth) and hidden (plowed fields, under a log, or away from the hide site).

(2) *Overhead cover.* In a semipermanent hide position, logs should be used as the base of the roof. The sniper team places a dust cover over the base (such as a poncho, layers of empty sandbags, or canvas), a layer of dirt, and a layer of gravel, if available. The team spreads another layer of dirt, and then adds camouflage. Due to the various materials, the roof is difficult to conceal if not countersunk.

(3) *Entrance.* To prevent detection, the sniper team should construct an entrance door sturdy enough to bear a man's weight.

(4) *Loopholes.* The construction of loopholes (Figure 4-9, page 4-18) requires care and practice to ensure they afford adequate fields of fire. Loopholes must be camouflaged by foliage or other material that blends with or is natural to the surroundings.

(5) *Approaches.* It is vital that the natural appearance of the ground remains unaltered and camouflage blends with the surroundings.

Construction time is wasted if the enemy observes a team entering the hide; therefore, approached must be concealed. Teams try to enter the hide during darkness, keeping movement to a minimum and adhering to trail discipline. In built-up areas, a secure and quiet approach is needed. Teams must avoid drawing attention to the mission and carefully plan movement. A possible ploy is to use a house search with sniper gear hidden among other gear. Sewers may be used for movement also.

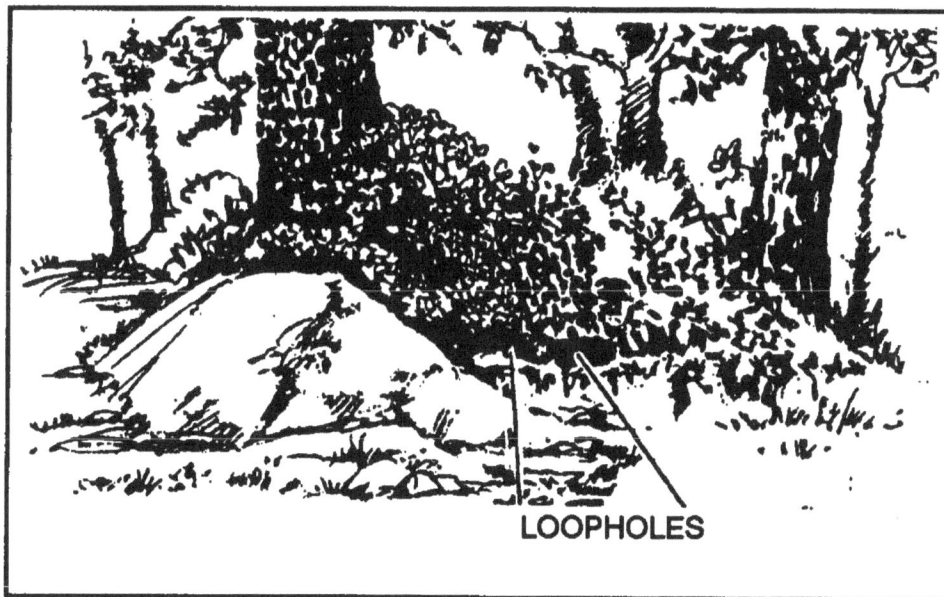

Figure 4-9. Loopholes in hide position.

c. **Hasty Position.** A hasty position is used when the sniper team is in a position for a short time and cannot construct a position due to the location of the enemy, or immediately assumes a position. The hasty position is characterized by the following

(1) *Advantages:*

(a) *Requires no construction* The sniper team uses what is available for cover and concealment.

(b) *Can be occupied in a short time.* As soon as a suitable position is found, the team need only prepare loopholes by moving small amounts of vegetation or by simply backing a few feet away from the vegetation that is already thereto conceal the weapon's muzzle blast.

(2) *Disadvantages:*

(a) *Affords no freedom of movement.* Any movement that is not slow and deliberate may result in the team being compromised.

(b) *Restricts observation of large areas.* This type of position is normally used to observe a specific target area (intersection, passage, or crossing).

(c) *Offers no protection from direct or indirect fires.*

(d) *Relies heavily on personal camouflage.* The team's only protection against detection is personal camouflage and the ability to use the available terrain.

(3) **Occupation time.** The team should not remain in this type of position longer than eight hours.

d. **Expedient Position.** When a sniper team is required to remain in position for a longer time than the hasty position can provide, an expedient position (Figure 4-10) should be constructed. The expedient position lowers the sniper's silhouette as low to the ground as possible, but it still allows him to fire and observe effectively. The expedient position is characterized by the following

(1) *Advantages:*

(a) *Requires little construction.* This position is constructed by digging a hole in the ground just large enough for the team and its equipment. Soil dug from this position can be placed in sandbags and used for building firing platforms.

(b) *Conceals most of the body and equipment.* The optics, rifles, and heads of the sniper team are the only items that are above ground level in this position.

(c) Provides some protection from direct fires due to its lower silhouette.

Figure 4-10. Expedient position.

(2) *Disadvantages:*

(a) *Affords little freedom of movement.* The team has more freedom of movement in this position than in the hasty position. Team members can lower their heads below ground level slowly to ensure a target indicator is not produced.

(b) *Allows little protection from indirect fires.* This position does not protect the team from shrapnel and debris falling into the position.

(c) *Exposes the head, weapons, and optics.* The team must rely heavily on the camouflaging of these exposed areas.

(3) *Construction time:* 1 to 3 hours (depending on the situation).

(4) *Occupation time:* 6 to 12 hours.

e. **Belly Hide.** The belly hide (Figure 4-11) is similar to the expedient position, but it has overhead cover that not only protects the team from the effects of indirect fires but also allows more freedom of movement. This position can be dugout under a tree, a rock, or any available object that provides overhead protection and a concealed entrance and exit. The belly hide is characterized by the following

(1) *Advantages:*

(a) *Allows some freedom of movement.* The darkened area inside this position allows the team to move freely. The team must remember to cover the entrance/exit door so outside light does not silhouette the team inside the position or give the position away.

(b) *Conceals all but the rifle barrel.* All equipment is inside the position except the rifle barrels. Depending on the room available to construct the position, the rifle barrels may also be inside.

(c) *Provides protection from direct and indirect fires.* The team should try to choose a position that has an object that will provide good overhead protection (rock tracked vehicle, rubble pile, and so forth), or prepare it in the same manner as overhead cover for other infantry positions.

(2) *Disadvantages:*

(a) *Requires extra construction time.*

(b) *Requires extra materials and tools.* Construction of overhead cover requires saws or axes, waterproof material, and so forth.

(c) *Has limited space.* The sniper team will have to lay in the belly hide without a lot of variation in body position due to limited space and design of the position.

(3) *Construction time:* 4 to 6 hours.

(4) *Occupation time:* 12 to 48 hours.

Figure 4-11. Belly hide position.

f. **Semipermanent Hide.** The semipermanent hide (Figure 4-12, page 4-22) is used mostly in defensive situations. This position requires additional equipment and personnel to construct. However, it allows sniper teams to remain in place for extended periods or to be relieved in place by other sniper teams. Like the belly hide, this position can be constructed by tunneling through a knoll or under natural objects already in place. The semipermanent hide is characterized by the following

(1) **Advantages:**

(a) *Offers total freedom of movement inside the position.* The team members can move about freely. They can stand, sit, or even lie down.

(b) *Protects against direct and indirect fires.* The sniper team should look for the same items as mentioned in the belly hide.

(c) *Is completely concealed.* Loopholes are the only part of the position that can be detected. They allow for the smallest exposure possible; yet they still allow the sniper and observer to view the target area. These loopholes should have a large diameter (10 to 14 inches) in the interior of the position and taper down to a smaller diameter (4 to 8 inches) on the outside of the position. A position may have more than two sets of loopholes if needed to cover large areas. The entrance/exit to the position must be covered to prevent light from entering and highlighting the loopholes. Loopholes that are not in use should be covered from the inside with a piece of canvas or suitable material.

(d) *Is easily maintained for extended periods.* This position allows the team to operate effectively for a longer period.

(2) *Disadvantages:*

(a) *Requires extra personnel and tools to construct.* This position requires extensive work and extra tools. It should not be constructed near the enemy. It should be constructed during darkness and be completed before dawn.

Figure 4-12. Semipermanent hide position.

(b) *Increases risk of detection.* Using a position for several days or having teams relieve each other in a position always increases the risk of detection.

(3) **Construction time:** 4 to 6 hours (4 personnel).

(4) **Occupation time:** 48 hours plus (relieved by other teams).

g. **Routines in Sniper Team positions.** Although the construction of positions may differ, the routines while in position are the same. The sniper and the observer should have a good firing platform. This gives the sniper a stable platform for the sniper weapon and the observer a platform for the optics. When rotating observation duties, the sniper weapon should remain in place, and the optics are handed from one member to the other. Sniper data book, observation logs, range cards, and the radio should be placed between the team where both members have easy access to them. A system of resting, eating, and latrine calls must be arranged between the team. All latrine calls should be done during darkness, if possible. A hole should be dug to conceal any traces of latrine calls.

4-13. POSITIONS IN URBAN TERRAIN

Positions in urban terrain are quite different than positions in the field. The sniper team normally has several places to choose. These can range from inside attics to street-level positions in basements. This type of terrain is ideal for a sniper, and a sniper team can stop an enemy's advance through its area of responsibility.

a. When constructing an urban position, the sniper team must be aware of the outside appearance of the structure. Shooting through loopholes in barricaded windows is preferred; the team must make sure all other windows are also barricaded. Building loopholes in other windows also provides more positions to engage targets. When building loopholes, the team should make them different shapes (not perfect squares or circles). Dummy loopholes also confuse the enemy. Positions in attics are also effective. The team removes the shingles and cuts out loopholes in the roof; however, they must make sure there are other shingles missing from the roof so the firing position loophole is not obvious.

(1) The sniper team should not locate the position against contrasting background or in prominent buildings that automatically draw attention. It must stay in the shadows while moving, observing, and engaging targets.

(2) The team must never fire close to a loophole. It should always back away from the hole as far as possible to hide the muzzle flash and to scatter the sound of the weapon when it fires. The snipers may be located in a different room than the loophole; however, they can make a hole through a wall to connect the rooms and fire from inside one room. The team must not fire continually from one position. (More than one position should be constructed if time and situation permit.) When constructing other positions, the team makes sure the target area can be observed. Sniper team positions should never be used by any personnel other than a sniper team.

b. Common sense and imagination are the sniper team's only limitation in the construction of urban hide positions. Urban hide positions that can be used are the room hide, crawl space hide, and rafter hide. The team constructs and occupies one of these positions or a variation thereof.

WARNING
WHEN MOVING THROUGH SEWERS, TEAMS MUST BE ALERT FOR BOOBY TRAPS AND POISONOUS GASES.

---

(1) *Room hide position.* In a room hide position, the sniper team uses an existing room and fires through a window or loophole (Figure 4-13). Weapon support may be achieved through the use of existing furniture-that is, desks or tables. When selecting a position, teams must notice both front and back window positions. To avoid. silhouetting, they may need to use a backdrop such as a dark-colored blanket, canvas, carpet, and a screen. Screens (common screening material) are important since they allow the sniper teams maximum observation and deny observation by the enemy. They must not. remove curtains; however, they can open windows or remove panes of glass. Remember, teams can randomly remove panes in other windows so the position is not obvious.

(2) *Crawl space hide position.* The sniper team builds a crawl space hide position in the space between floors in multistory buildings (Figure 4-14). Loopholes are difficult to construct, but a damaged building helps considerably. Escape routes can be holes knocked into the floor or ceiling. Carpet or furniture placed over escape holes or replaced ceiling tiles will conceal them until needed.

Figure 4-13. Room hide position.

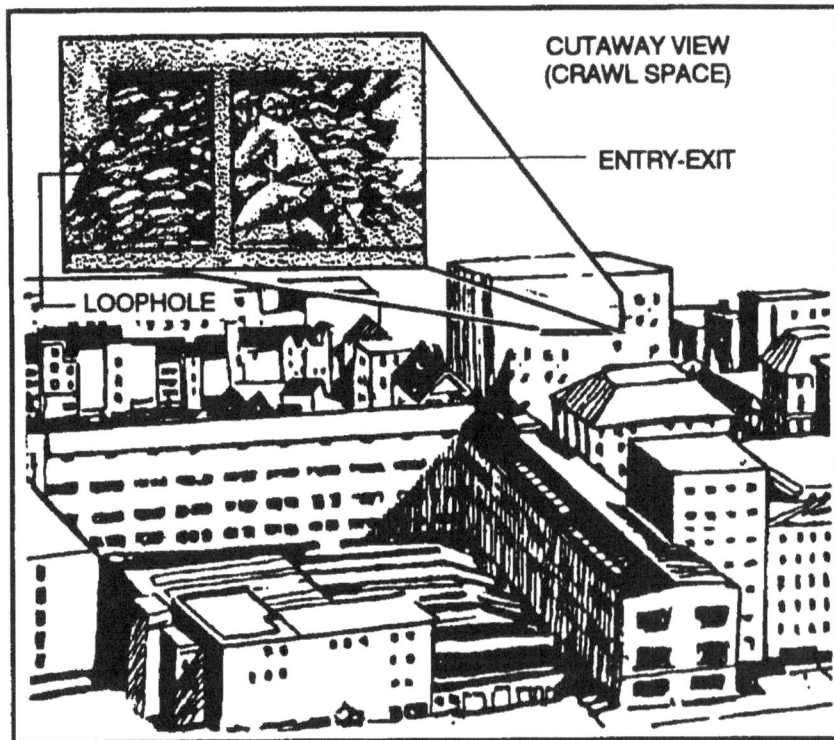

Figure 4-14. Crawl space hide position.

(3) *Rafter hide position.* The sniper team constructs a rafter hide position in the attic of an A-frame-type building. These buildings normally have shingled roofs (A and B, Figure 4-15). Firing from inside the attic around a chimney or other structure helps prevent enemy observation and fire.

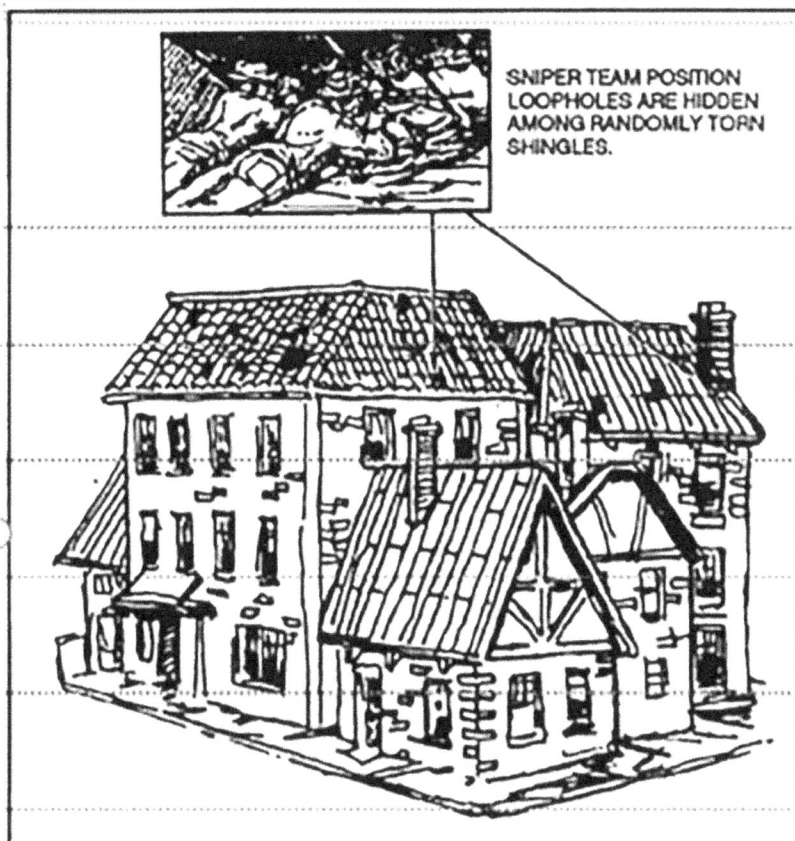

SNIPER TEAM POSITION LOOPHOLES ARE HIDDEN AMONG RANDOMLY TORN SHINGLES.

Figure 4-15. Rafter hide positions.

c. Sniper teams use the technique best suited for the urban hide position.

(1) The second floor of a building is usually the best location for the position. It presents minimal dead space but provides the team more protection since passersby cannot easily spot it.

(2) Normally, a window is the best viewing aperture/loophole.

(a) If the window is dirty, do not clean it for better viewing.

(b) If curtains are prevalent in the area, do not remove those in the position. Lace or net-type curtains can be seen through from the inside, but they are difficult to see through from the outside.

(c) If strong winds blow the curtains open, staple, tack, or weight them.

(d) Firing a round through a curtain has little effect on accuracy however, ensure the muzzle is far enough away to avoid muzzle blast.

(e) When area routine indicates open curtains, follow suit. Set up well away from the loophole; however, ensure effective coverage of the assigned target area.

(3) Firing through glass should be avoided since more than one shot may be required. The team considers the following options:

(a) Break or open several windows throughout the position before occupation. This can be done during the reconnaissance phase of the operation; however, avoid drawing attention to the area.

(b) Remove or replace panes of glass with plastic.

(4) Other loopholes/ viewing apertures are nearly unlimited.
- Battle damage.
- Drilled holes (hand drill).
- Brick removal.
- Loose boards/ derelict houses.

(5) Positions can also beset up in attics or between the ceiling and roof. (See rafter hide positions.)
- Gable ends close to the eaves (shadow adding to concealment).
- Battle damage to gables and or roof.
- Loose or removed tiles, shingles, or slates.
- Skylights.

(6) The sniper makes sure the bullet clears the loophole. The muzzle must be far enough from the loophole to ensure the bullet's path is not in line with the bottom of the loophole.

(7) Front drops, usually netting, may have to be changed (if the situation permits) from dark to light colors at BMNT/ EENT due to sunlight or lack of sunlight into the position.

(8) If the site is not multiroomed, partitions can be made by hanging blankets or nets to separate the operating area from the rest/ administrative area.

(9) If sandbags are required, they can be filled and carried inside of rucksacks or can be filled in the basement, depending on the situation/location of the position site.

(10) Always plan an escape route that leads to the objective rally point. When forced to vacate the position, the team meets the security element at the ORP. Normally, the team will not be able to leave from the same point at which it gained access; therefore, a separate escape point may be required in emergency situations. The team must consider windows (other than the viewing apertures); anchored ropes to climb down buildings, or a small, preset explosive charge situated on a wall or floor for access into adjoining rooms, buildings, or the outside.

(11) The type of uniform or camouflage to be worn by the team will be dictated by the situation, how they are employed, and area of operation. The following applies:

(a) Most often, the BDU and required equipment are worn.

(b) Urban-camouflaged uniforms can be made or purchased. Urban areas vary in color (mostly gray [cinder block]; red [brick]; white [marble]; black [granite]; or stucco, clay, or wood). Regardless of area color, uniforms should include angular-line patterns.

(c) When necessary, most woodland-patterned BDUs can be worn inside out as they are a gray or green-gray color underneath.

(d) Soft-soled shoes or boots are the preferred footwear in the urban environment.

(e) Civilian clothing can be worn (native/host country populace).

(f) Tradesmen's or construction worker's uniforms and accessories can be used.

Section IV
OBSERVATION

Throughout history, battles have been won and nations conquered based on an accurate accounting and description of the opposing forces strength, equipment, and location. As the sniper team performs the secondary mission of collecting and reporting battlefield intelligence, the commander can act, rather than react. The purpose of observation is to gather facts and to provide information for a specific intent. Observation uses all of the sniper team's five senses but often depends on sight and hearing. For example, the sniper team is issued a PIR or OIR for a specific mission. Information gathered by the sniper team is reported, analyzed, and processed into intelligence reports. The sniper team's success depends upon its powers of observation. In addition to the sniperscope, the sniper team has an observation telescope, binoculars, night vision sight, and night vision goggles to enhance its ability to observe and engage targets. Team members must relieve each other when using this equipment since prolonged use can cause eye fatigue, greatly reducing the effectiveness of observation. Team members rotate periods of observation. During daylight, observation should be limited to 10 minutes followed by a 10-minute rest. When using night vision devices, the observer should limit his initial period of viewing to 10 minutes followed by a 10-minute rest. After several periods of viewing, he can extend the viewing period to 15 minutes and then a 15-minute rest.

4-14. HASTY AND DETAILED SEARCHES

While observing a target area, the sniper team alternately conducts two types of visual searches: hasty and detailed.

a. A hasty search is the first phase of observing a target area. The observer conducts a hasty search immediately after the team occupies the firing position. A hasty search consists of quick glances with binoculars at specific points, terrain features, or other areas that could conceal the enemy. The observer views the area closest to the team's position first since it could pose the most immediate threat. The observer then searches farther out until the entire target area has been searched. When the observer sees or suspects a target, he uses an M49 observation telescope for a detailed view of the target area. The telescope should not be used to search the area because its narrow field of view would take much longer to cover an area; plus, its stronger magnification can cause eye fatigue sooner than the binoculars.

b. After a hasty search has been completed, the observer then conducts a detailed search of the area. A detailed search is a closer, more thorough search of the target area, using 180-degree area or sweeps, 50 meters in depth, and overlapping each previous sweep at least 10 meters to ensure the entire area has been observed (Figure 4-16, page 4-30). Like the hasty search, the observer begins by searching the area closest to the sniper team position.

c. This cycle of a hasty search followed by a detailed search should be repeated three or four times. This allows the sniper team to become accustomed to the area; plus, the team will look closer at various points with each consecutive pass over the area. After the initial searches, the observer should view the area, using a combination of both hasty and detailed searches. While the observer conducts the initial searches of the area, the sniper should record prominent features, reference points, and distances on a range card. The team members should alternate the task of observing the area about every 30 minutes.

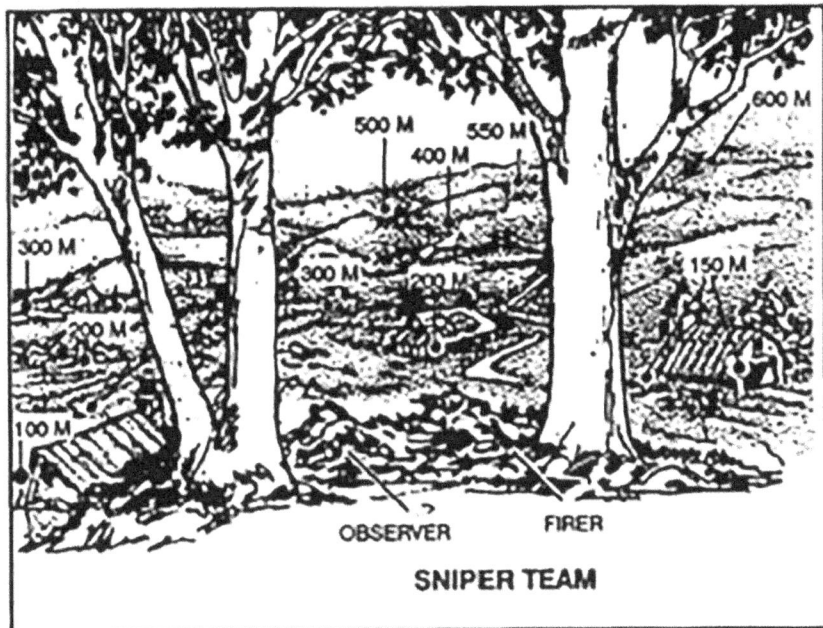

Figure 4-16. Detailed search.

4-15. ELEMENTS OF OBSERVATION

The four elements in the process of observation include awareness, understanding, recording, and response. Each of these elements may be accomplished as a separate process or accomplished at the same time.

a. Awareness. Awareness is being consciously attuned to a specific fact. A sniper team must always be aware of the surroundings and take nothing for granted. The team also considers certain elements that influence and distort awareness.

(1) An object's size and shape can be misinterpreted if viewed incompletely or inaccurately.

(2) Distractions degrade the quality of observations.

(3) Active participation or degree of interest can diminish toward the event.

(4) Physical abilities (five senses) have limitations.

(5) Environmental changes affect accuracy.

(6) Imagination may cause possible exaggerations or inaccuracy.

b. Understanding. Understanding is derived from education, training, practice, and experience. It enhances the sniper team's knowledge about what should be observed, broadens its ability to view and consider all aspects, and aids in its evaluation of information.

c. **Recording.** Recording is the ability to save and recall what was observed. Usually, the sniper team has mechanical aids, such as writing utensils, sniper data book, sketch kits, tape recorders, and cameras, to support the recording of events; however, the most accessible method is memory. The ability to record, retain, and recall depends on the team's mental capacity (and alertness) and ability to recognize what is essential to record. Added factors that affect recording include

(1) The amount of training and practice in observation.

(2) Skill gained through experience.

(3) Similarity of previous incidents.

(4) Time interval between observing and recording.

(5) The ability to understand or convey messages through oral or other communications.

d. **Response.** Response is the sniper team's action toward information. It may be as simple as recording events in a sniper data book, making a communications call, or firing a well-aimed shot.

NOTE: See Chapter 9 for discussion on the keep-in-memory (KIM) game.

4-16. TWILIGHT TECHNIQUES

Twilight induces a false sense of security, and the sniper team must be extremely cautious. The enemy is also prone to carelessness and more likely to expose himself at twilight. During twilight, snipers should be alert to OP locations for future reference. The M3A telescope reticle is still visible and capable of accurate fire 30 minutes before BMNT and 30 minutes after EENT.

4-17. NIGHT TECHNIQUES

Without night vision devices, the sniper team must depend upon eyesight. Regardless of night brightness, the human eye cannot function at night with daylight precision. For maximum effectiveness, the sniper team must apply the following principles of night vision:

a. **Night Adaptation.** The sniper team should wear sunglasses or red-lensed goggles in lighted areas before departing on a mission. After departure, the team makes a darkness adaptation and listening halt for 30 minutes.

b. **Off-Center Vision.** In dim light, an object under direct focus blurs, appears to change, and sometimes fades out entirely. However, when the eyes are focused at different points, about 5 to 10 degrees away from an object, peripheral vision provides a true picture. This allows the light-sensitive portion of the eye, that not used during the day, to be used.

c. Factors Affecting Night Vision. The sniper team has control over the following night vision factors:

(1) Lack of vitamin A impairs night vision. However, an overdose of vitamin A will not improve night vision capability.

(2) Colds, fatigue, narcotics, headaches, smoking, and alcohol reduce night vision.

(3) Exposure to bright light degrades night vision and requires a readaption to darkness.

4-18. ILLUMINATION AIDS

The sniper team may occasionally have artificial illumination for observing and firing. Examples are artillery illumination fire, campfires, or lighted buildings.

a. Artillery Illumination Fire. The M301A2 illuminating cartridge provides 50,000 candlepower.

b. Campfires. Poorly disciplined enemy soldiers may use campfires, or fires may be created by battlefield damage. These opportunities give the sniper enough illumination for aiming.

c. Lighted Buildings. The sniper can use lighted buildings to eliminate occupants of the building or personnel in the immediate area of the light source.

Section V
TARGET DETECTION AND SELECTION

Recording the type and location of targets in the area helps the sniper team to determine engageable targets. The sniper team must select key targets that will do the greatest harm to the enemy in a given situation. It must also consider the use of indirect fire on targets. Some targets, due to their size or location, may be better engaged with indirect fire.

4-19. TARGET INDEXING

To index targets, the sniper team uses the prepared range card for a reference since it can greatly reduce the engagement time. When indexing a target to the sniper, the observer locates a prominent terrain feature near the target. He indicates this feature and any other information to the sniper to assist in finding the target. Information between team members varies with the situation. The observer may sound like an FO giving a call for fire to an FDC depending on the condition of the battlefield and the total number of possible targets from which to choose.

a. **Purpose.** The sniper team indexes targets for the following reasons:

(1) Sniper teams may occupy an FFP in advance of an attack to locate, index, and record target locations; and to decide on the priority of targets.

(2) Indiscriminate firing may alert more valuable and closer enemy targets.

(3) Engagement of a distant target may result in disclosure of the FFP to a closer enemy.

(4) A system is needed to remember location if several targets are sighted at the same time.

b. **Considerations.** The sniper team must consider the following factors when indexing targets:

(1) *Exposure times.* Moving targets may expose themselves for only a short time The sniper team must note the point of disappearance of each target, if possible, before engagement. By doing so, the team may be able to take several targets under fire in rapid succession.

(2) *Number of targets.* If several targets appear and disappear at the same time, the point of disappearance of each is hard to determine; therefore, sniper teams concentrate on the most important targets.

(3) *Spacing/distance between targets.* The greater the distance between targets, the harder it is to see their movement. In such cases, the team should locate and engage the nearest targets.

(4) *Evacuation of aiming points.* Targets that disappear behind good aiming points are easily recorded and remembered, targets with poor aiming points are easily lost. Assuming that two such targets are of equal value and danger, the team should engage the more dangerous aiming point target first.

c. **Determination of Location of Hidden Fires.** When using the *crack-thump method,* the team listens for the crack of the round and the thump of the weapon being fired. By using this method, the sniper can obtain both a direction and a distance.

(1) *Distance to firer. The* time difference between the crack and the thump can be converted into an approximate range. A one-second lapse between the two is about 600 yards with most calibers; a one-half-second lapse is about 300 yards.

(2) *Location of firer. By* observing in the direction of the thump and near the predetermined range, the sniper team has a good chance of seeing the enemy's muzzle flash or blast from subsequent shots.

(3) *Limitations.*The crack-thump method has the following limitations

(a) Isolating the crack and thump is difficult when many shots are being fired.

(b) Mountainous areas, tall buildings, and so forth cause echoes and make this method ineffective.

d. **Shot-Hole Analysis.** Locating two or more shot holes in trees, walls, dummy heads, and so forth may make it possible to determine the direction of the shots. The team can use the dummy-head pencil method and triangulate on the enemy sniper's position. However, this method only works if all shots come from the same position.

4-20. TARGET SELECTION

Target selection may be forced upon the sniper team. A target moving rapidly may be lost while obtaining positive identification. The sniper team considers any enemy threatening its position as a high-value target. When selecting key targets, the team must consider the following factors:

a. **Threat to the Sniper Team.** The sniper team must consider the danger the target presents. This can be an immediate threat, such as an enemy element walking upon its position, or a future threat, such as enemy snipers or dog tracking teams.

b. **Probability of First-Round Hit.** The sniper team must determine the chances of hitting the target with the first shot by considering the following:

- Distance to the target.
- Direction and velocity of the wind.
- Visibility of the target area.
- Amount of the target that is exposed.
- Amount of time the target is exposed.
- Speed and direction of target movement.

c. **Certainty of Target's Identity.** The sniper team must be reasonably certain that the target it is considering is the key target.

d. **Target Effect on the Enemy.** The sniper team must consider what effect the elimination of the target will have on the enemy's fighting ability It must determine that the target is the one available target that will cause the greatest harm to the enemy.

e. **Enemy Reaction to Sniper Fire.** The sniper team must consider what the enemy will do once the shot has been fired. The team must be prepared for such actions as immediate suppression by indirect fires and enemy sweeps of the area.

f. **Effect on the Overall Mission.** The sniper team must consider how the engagement will affect the overall mission. The mission may be one of intelligence gathering for a certain period. Firing will not only alert the enemy to a team's presence, but it may also terminate the mission if the team has to move from its position as a result of the engagement.

4-21. KEY TARGETS

Key personnel targets can be identified by actions or mannerisms, by positions within formations, by rank or insignias, and or by equipment being worn or carried. Key targets can also include weapon systems and equipment. Examples of key targets areas follows:

a. **Snipers.** Snipers are the number one target of a sniper team. The enemy sniper not only poses a threat to friendly forces, but he is also the natural enemy of the sniper. The fleeting nature of a sniper is reason enough to engage him because he may never be seen again.

b. **Dog Tracking Teams.** Dog tracking teams pose a great threat to sniper teams and other special teams that may be working in the area. It is hard to fool a trained dog. When engaging a dog tracking team, the sniper should engage the dog's handler first. This confuses the dog, and other team members may not be able to control it.

c. **Scouts.** Scouts are keen observers and provide valuable information about friendly units. This plus their ability to control indirect fires make them dangerous on the battlefield. Scouts must be eliminated.

d. **Officers.** Officers are another key target of the sniper team. Losing key officers in some forces is such a major disruption to the operation that forces may not be able to coordinate for hours.

e. **Noncommissioned Officers.** Losing NCOs not only affects the operation of a unit but also affects the morale of lower ranking personnel,

f. **Vehicle Commanders and Drivers.** Many vehicles are rendered useless without a commander or driver.

g. **Communications Personnel.** In some forces, only highly trained personnel know how to operate various types of radios. Eliminating these personnel can be a serious blow to the enemy's communication network.

h. **Weapon Crews.** Eliminating weapon crews reduces the amount of fire on friendly troops.

i. **Optics on Vehicles.** Personnel who are in closed vehicles are limited to viewing through optics. The sniper can blind a vehicle by damaging these optic systems.

j. **Communication and Radar Equipment.** The right shot in the right place can completely ruin a tactically valuable radar or communication system. Also, only highly trained personnel may attempt to repair these systems in place. Eliminating these personnel may impair the enemy's ability to perform field repair.

k. **Weapon Systems.** Many high-technology weapons, especially computer-guided systems, can be rendered useless by one well-placed round in the guidance controller of the system.

Section VI
RANGE ESTIMATION

A sniper team is required to accurately determine distance, to properly adjust elevation on the sniper weapon system, and to prepare topographical sketches or range cards. Because of this, the team has to be skilled in various range estimation techniques.

4-22. FACTORS AFFECTING RANGE ESTIMATION

Three factors affect range estimation: nature of the target, nature of the terrain, and light conditions.

a. **Nature of the Target.**

(1) An object of regular outline, such as a house, appears closer than one of irregular outline, such as a clump of trees.

(2) A target that contrasts with its background appears to be closer than it actually is.

(3) A partly exposed target appears more distant than it actually is.

b. **Nature of the Terrain.**

(1) As the observer's eye follows the contour of the terrain, he tends to overestimate distant targets.

(2) Observing over smooth terrain, such as sand, water, or snow, causes the observer to underestimate distant targets.

(3) Looking downhill, the target appears farther away.

(4) Looking uphill, the target appears closer.

c. **Light Conditions.**

(1) The more clearly a target can be seen, the closer it appears.

(2) When the sun is behind the observer, the target appears to be closer.

(3) When the sun is behind the target, the target is more difficult to see and appears to be farther away.

4-23. RANGE ESTIMATION METHODS

Sniper teams use range estimation methods to determine distance between their position and the target.

a. **Paper-Strip Method.** The paper-strip method (Figure 4-17) is useful when determining longer distances (1,000 meters plus). When using this method, the sniper places the edge of a strip of paper on the map and ensures it is long enough to reach between the two points. Then he pencils in a tick mark on the paper at the team position and another at the distant location. He places the paper on the map's bar scale, located at the bottom center of the map, and aligns the left tick mark with the 0 on the scale. Then he reads to the right to the second mark and notes the corresponding distance represented between the two marks.

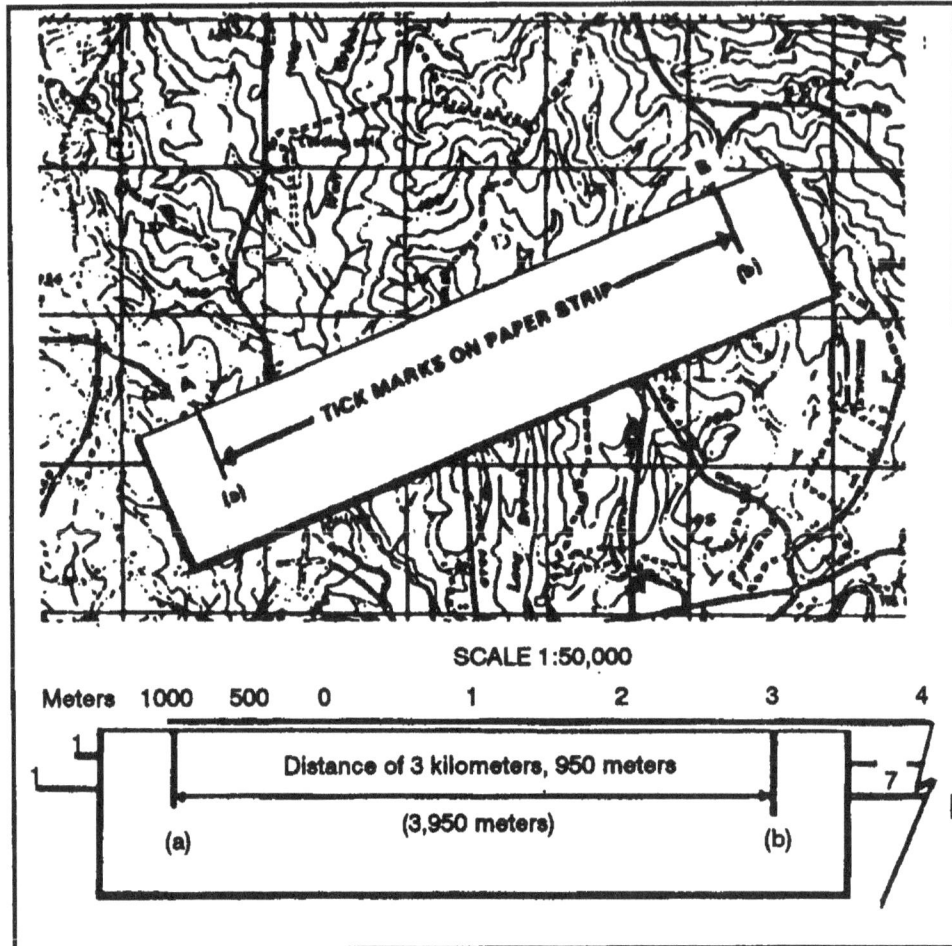

Figure 4-17. Paper-strip method.

b. **100-Meter-Unit-of-Measure Method.** To use this method (Figure 4-18, page 4-38), the sniper team must be able to visualize a distance of 100 meters on the ground. For ranges up to 500 meters, the team determines the number of 100-meter increments between the two objects it wishes to measure. Beyond 500 meters, it must select a point halfway to the object and determine the number of 100-meter increments to the halfway point, then double it to find the range to the object.

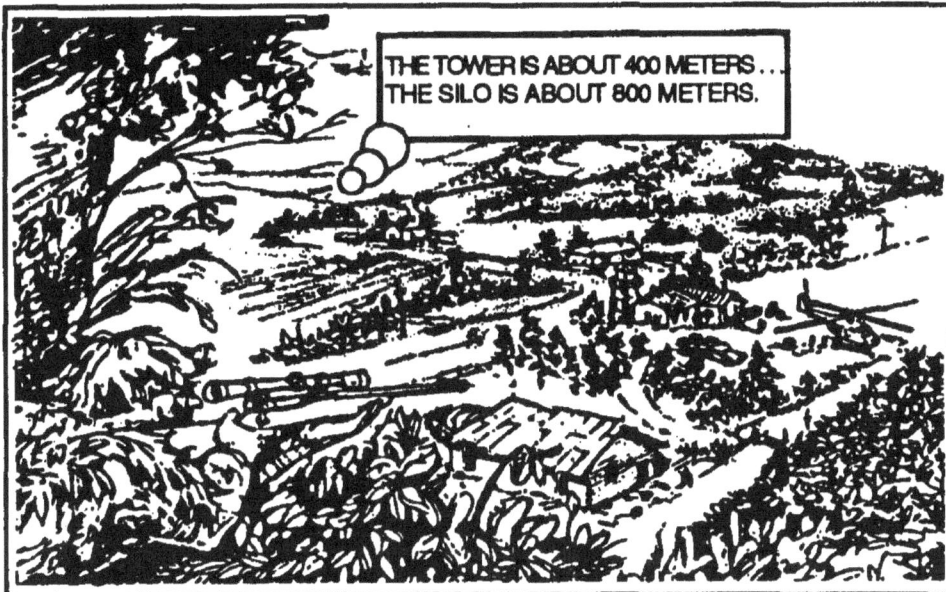

Figure 4-18. 100-meter-unit-of-measure method.

c. **Appearance-of-Object Method.** This method is a means of determining range by the size and other characteristic details of the object. To use the appearance-of-object method with any degree of accuracy, the sniper team must be familiar with the characteristic details of the objects as they appear at various ranges.

d. **Bracketing Method.** Using this method, the sniper team assumes that the target is no more than X meters but no less than Y meters away. An average of X and Y will be the estimate of the distance to the target.

e. **Range-Card Method.** The sniper team an also use a range card to quickly determine ranges throughout the target area. Once a target is seen, the team determines where it is located on the card and then reads the proper range to the target.

f. **Mil-Relation Formula.** The mil-relation formula is the preferred method of range estimation. This method uses a mil-scale reticle located in the M19 binoculars (Figure 4-19) or in the M3A sniperscope (Figure 4-20). The team must know the target size in inches or meters. Once the target size is known, the team then compares the target size to the mil-scale reticle and uses the following formula:

$$\frac{\text{Size of target in meters} \times 1{,}000}{\text{Size of object in mils}} = \text{Range to target in meters}$$

(To convert inches to meters, multiply the number of inches by .0254.)

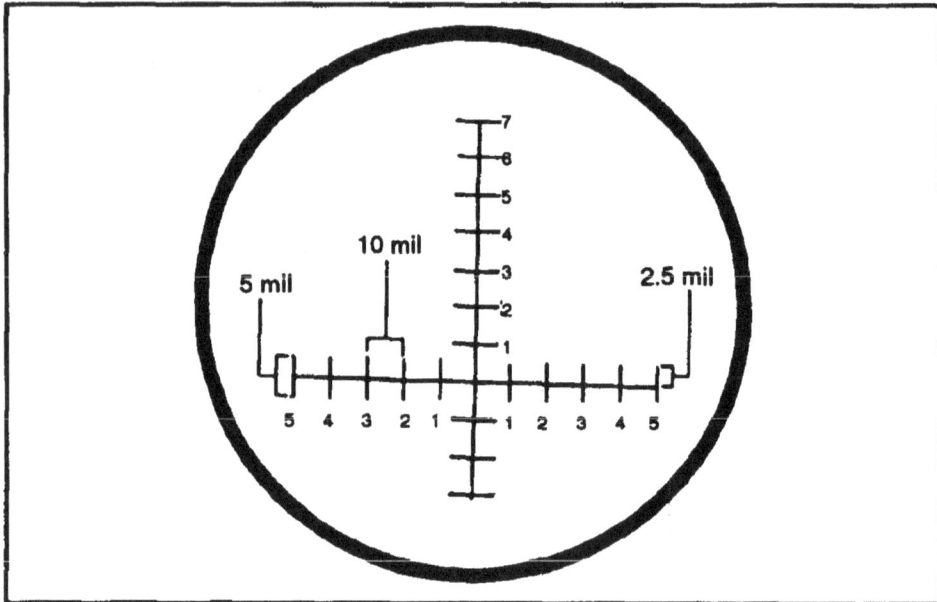

Figure 4-19. M19 mil-scale reticle.

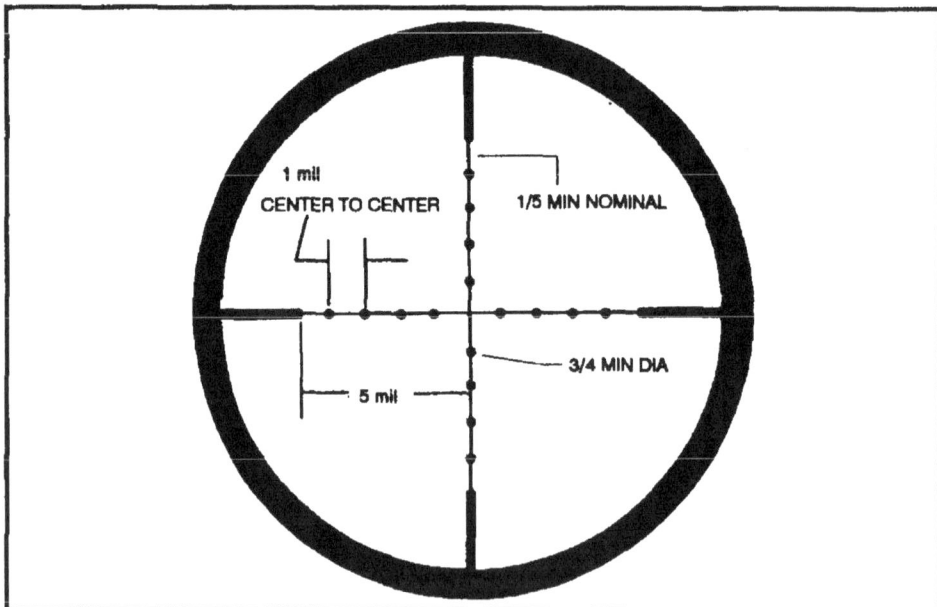

Figure 4-20. M3A mil-scale reticle.

g. **Combination Method.** In a combat environment, perfect conditions rarely exist. Therefore, only one method of range estimation may not be enough for the team's specific mission. Terrain with much dead space limits the accuracy of the 100-meter method. Poor visibility limits the use of the appearance-of-object method. However, by using a combination of two or more methods to determine an unknown range, an experienced sniper team should arrive at an estimated range close to the true range.

4-24. LASER RANGE FINDER

When the sniper team has access to a laser observation set, AN/ GVS-5, the set should always be used. It can provide the sniper team range to a specific target with great accuracy. When aiming the laser at a specific target, the sniper should support it much the same as his weapon to ensure accuracy. If the target is too small, aiming the laser at a larger object near the target will suffice (that is, a building, vehicle, tree, or terrain feature.)

4-25. ESTIMATION GUIDELINES

If mirage is too heavy to distinguish the bottom of a target, it should be halved.

EXAMPLE

When the target is estimated to be 70 inches high, divide the height into one-half. Use the following mil-relation formula:

$$\frac{35 \text{ inches} \times .0254 \times 1,000}{\text{Size of target in mils}} = \text{Range to target in meters}$$

By using this technique, estimate range to targets that are only partly visible. Such as:

The normal distance from the breastbone to the top of the head is 19 inches.

$$\frac{19 \text{ inches} \times .0254 \times 1,000}{\text{Size of target in mils}} = \text{Range to target in meters}$$

OR

Normal height of the human head is 10 inches.

$$\frac{10 \text{ inches} \times .0254 \times 1,000}{\text{Size of target in mils}} = \text{Range to target in meters}$$

This example may prove to be of specific use when facing an enemy entrenched in bunkers or in dense vegetation.

a. The sniper team should keep a sniper data book complete with measurements.

(1) *Vehicles.*
- Height of road wheels.
- Vehicle dimensions.
- Length of main gun tubes on tanks.
- Lengths/ sizes of different weapon systems.

(2) *Average height of human targets in area of operation.*

(3) *Urban environment.*
- Average size of doorways.
- Average size of windows.
- Average width of streets and lanes (average width of a paved road in the United States is 10 feet).
- Height of soda machines.

b. As the sniper team develops a sniper data book, all measurements are converted into constants and computed with different mil readings. An example of this is Table 4-1, which has already been computed for immediate use. This table should be incorporated into the sniper data book

TABLE FOR 6-FOOT MAN		
HEIGHT IN MILS	STANDING	SITTING/ KNEELING
1	2000	1000
1.5	1333	666
2	1000	500
2.5	800	400
3	666	333
3.5	571	286
4	500	250
4.5	444	222
5	400	200
5.5	364	182
6	333	167
6.5	308	154
7	286	143

TABLE FOR 5-FOOT 6-INCH MAN		
HEIGHT IN MILS	STANDING	SITTING/ KNEELING
1	1800	900
1.5	1200	600
2	900	450
2.5	750	375
3	600	300
3.5	514	257
4	450	225
4.5	400	200
5	360	180
5.5	327	164
6	300	150
6.5	277	139

Table 4-1. Range estimation table.

Section VII
INFORMATION RECORDS

The secondary mission of the sniper team is the collection and reporting of information. To accomplish this, the sniper team not only needs to be keen observers, but it also must accurately relay the information it has observed. To record this information, the team uses the sniper data book, which contains a range card, a military sketch, and an observation log.

4-26. RANGE CARD

The range card represents the target area drawn as seen from above with annotations indicating distances throughout the target area. Information is recorded on DA Form 5787-R (Sniper's Range Card) (Figure 4-21). (A blank copy of this form is located in the back of this publication for local reproduction.) The range card provides the sniper team with a quick-range reference and a means to record target locations, since it has preprinted range rings on it. These cards can be divided into sectors by using dashed lines. This provides the team members with a quick reference when locating targets-for example: "The intersection in sector A." A range card can be prepared on any paper the team has available. The sniper team position and distances to prominent objects and terrain features are drawn on the card. There is not a set maximum range on the range card, because the team may also label any indirect fire targets on its range card. Information contained on range cards includes:

    a. Name, rank, SSN, and unit.

    b. Method of obtaining range.

    c. Left and right limits of engageable area.

    d. Major terrain features, roads, and structures.

    e. Ranges, elevation, and windage needed at various distances.

    f. Distances throughout the area.

    g. Temperature and wind. (Cross out previous entry whenever temperature, wind direction, or wind velocity changes.)

    h. Target reference points (azimuth, distance, and description).

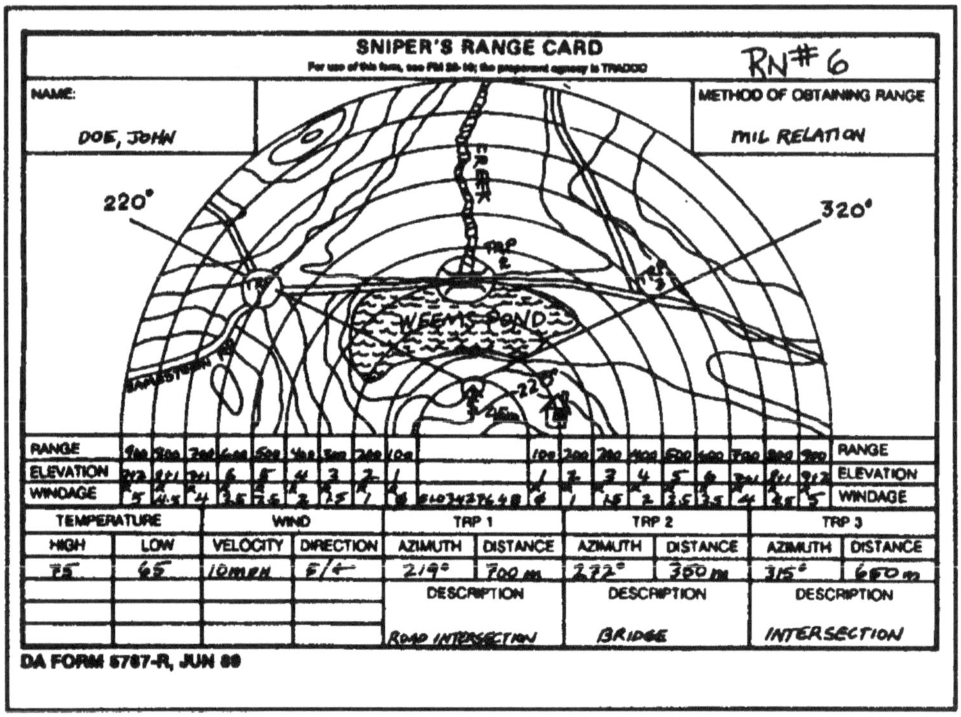

Figure 4-21. Example of completed DA Form 5787-R.

4-27. MILITARY SKETCH

DA Form 5788-R (Military Sketch) is used to record information about a general area, terrain features, or man-made structures that are not shown on a map. Military sketches provide intelligence sections a detailed, on-the-ground view of an area or object that is otherwise unobtainable. These sketches not only let the viewer see the area in different perspectives but also provide detail such as type of fences, number of telephone wires, present depth of streams, and so forth. There are two types of military sketches as stated in FM 21-26 panoramic sketches and topographic sketches. Information is recorded on DA Form 5788-R. (A blank copy of this form is located in the back of this publication for local reproduction.)

a. **Panoramic.** A panoramic sketch (Figure 4-22, page 4-44) is a representation of an area or object drawn to scale as seen from the sniper team's perspective. It shows details about a specific area or a man-made structure. Information considered in a panoramic sketch includes the following:

(1) Name, rank, SSN, and unit.

(2) Remarks section (two).

(3) Sketch name.
(4) Grid coordinates of sniper team's position.
(5) Weather.
(6) Magnetic azimuth through the center of sketch.
(7) Sketch number and scale of sketch.
(8) Date and time.

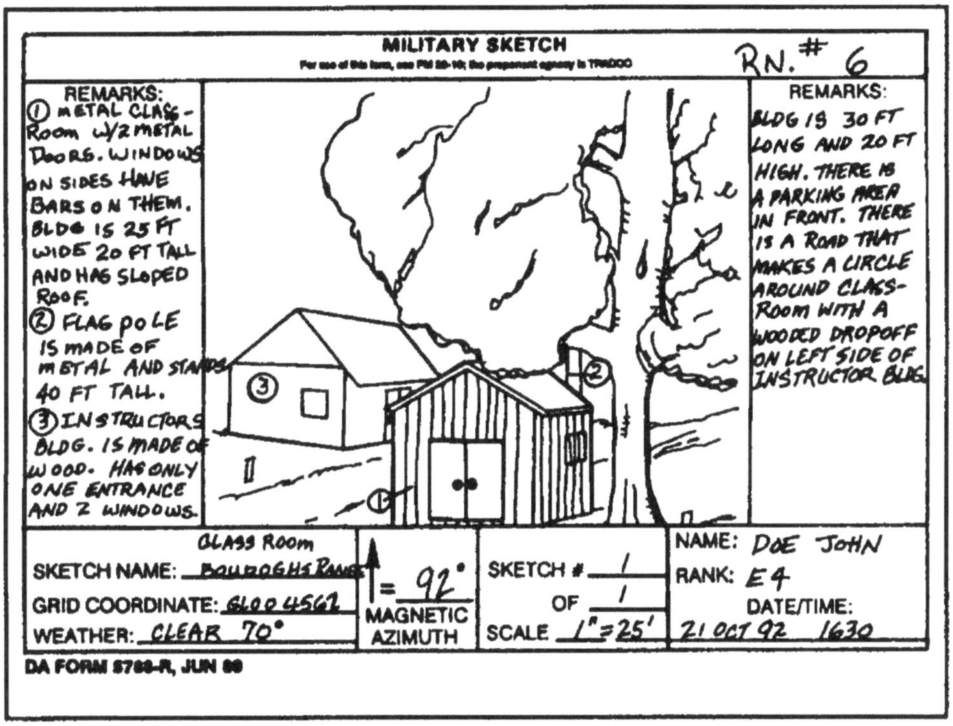

**Figure 4-22. Example of completed DA Form 5788-R
for panoramic sketch.**

b. **Topographic Sketch.** A topographic sketch (Figure 4-23) is a topographic representation of an area drawn to scale as seen from above. It provides the sniper team with a method for describing large areas while showing reliable distance and azimuths between major features. This type of sketch is useful in describing road systems, flow of streams/ rivers, or locations of natural and man-made obstacles. 'he field sketch can also be used as an overlay on the range card. Information contained in a field sketch includes the following

(1) Grid coordinates of the sniper team's position.
(2) Name, rank, SSN, and unit.
(3) Remarks.
(4) Sketch name.

(5) Grid coordinates.
(6) Weather.
(7) Magnetic azimuth.
(8) Sketch number and scale.
(9) Date and time.

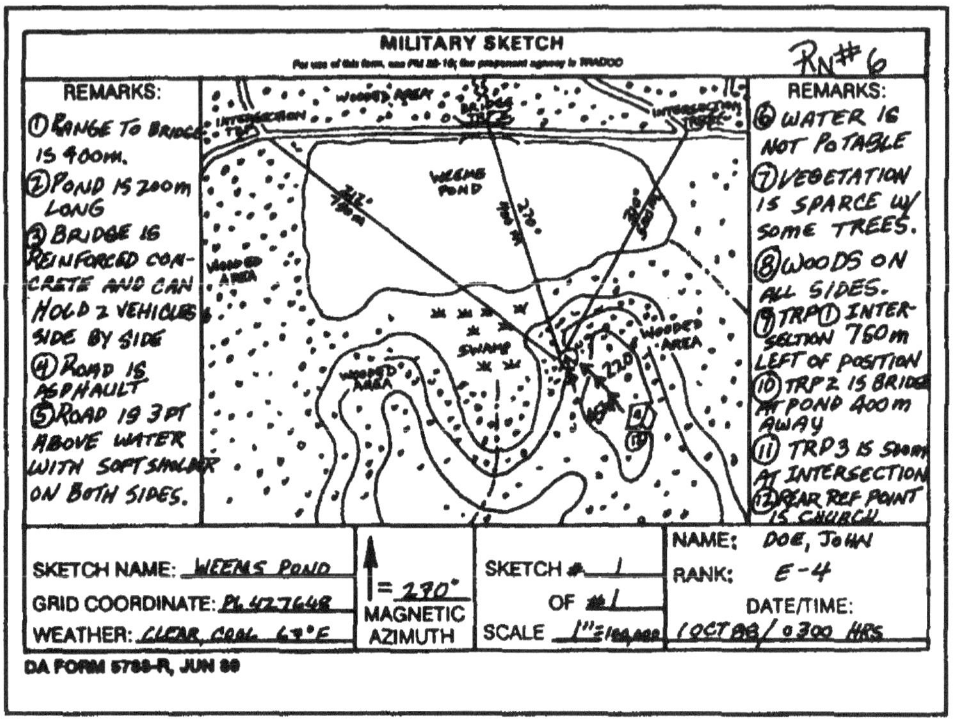

Figure 4-23. Example of completed DA Form 5788-R for topographic sketch.

c. **Guidelines for Drawing Sketches.** As with all drawings, artistic skill is an asset, but satisfactory sketches can be drawn by anyone with practice. The following are guidelines when drawing sketches:

(1) *Work from the whole to the part.* First determine the boundaries of the sketch. Then sketch the larger objects such as hills, mountains, or outlines of large buildings. After drawing the large objects in the sketch, start drawing the smaller details.

(2) *Use common shapes to show common objects.* Do not sketch each individual tree, hedgerow, or wood line exactly. Use common shapes to show these types of objects. Do not concentrate on the fine details unless they are of tactical importance.

(3) *Draw in perspective; use vanishing points.* Try to draw sketches in perspective. To do this, recognize the vanishing points of the area to

be sketched. Parallel lines on the ground that are horizontal vanish at a point on the horizon (Figure 4-24). Parallel lines on the ground that slope downward away from the observer vanish at a point below the horizon. Parallel lines on the ground that slope upward, away from the observer vanish at a point above the horizon. Parallel lines that recede to the right vanish on the right and those that recede to the left vanish on the left (Figure 4-24).

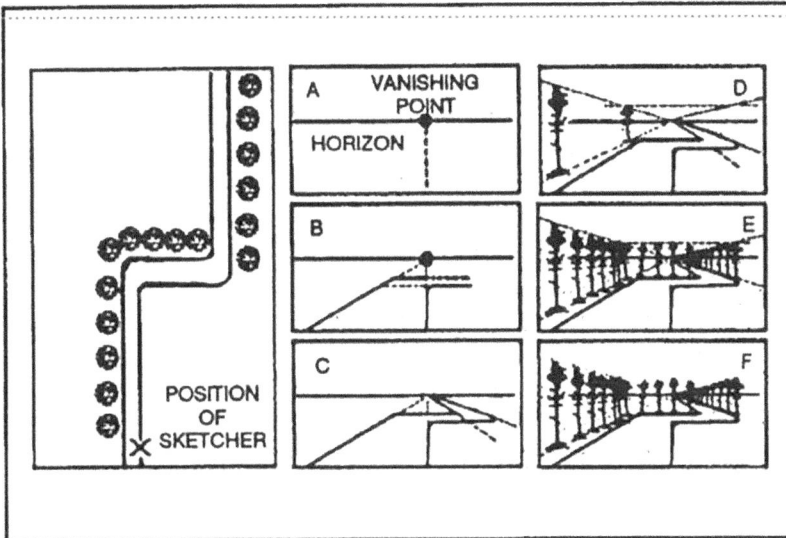

Figure 4-24. Vanishing points.

4-28. SNIPER DATA BOOK

The sniper data book is a written, chronological record of all activities and events that take place in a sniper team's area. It is used with military sketches and range cards; this combination not only gives commanders and intelligence personnel information about the appearance of the area, but it also provides an accurate record of the activity in the area. Information is recorded on DA Form 5786-R (Sniper's Observation Log) (Figure 4-25). (A blank copy of this form is in the back of this publication for local reproduction.) Information in the observation log includes: (Completion of this form is self-explanatory.)

    a. Sheet number and number of total sheets.

    b. Observer's name, rank, SSN, and unit.

    c. Date and time of observation and visibility.

d. Grid coordinates of the sniper team's position.
e. Series number, time, and grid coordinates of each event.
f. The event that has taken place.
g. Action taken and remarks.

SNIPER'S OBSERVATION LOG
For use of this form, see FM 23-10; the proponent agency is TRADOC

SHEET____OF____SHEETS

ORIGINATOR: DOE, JOHN R | DATE/TIME: 1 OCT 92 | LOCATION: GL03427648

SERIAL	TIME	GRID COORDINATE	EVENT	ACTIONS OR REMARKS
1	0300	GL03427648	OCCUPIED POSITION	OBSERVATION
2	0340	SAME	PFC JUDSON RESTED	NONE
3	0430	SAME	PFC JUDSON ASSUMED OBSERVATION	I RESTED
4	0530	SAME	BOTH OF US AWAKE	NONE
5	0630	SAME	PREPARED RANGE CARD AND TOPOGRAPHIC SKETCH	LIGHT ENOUGH TO SEE
6	0655	GL034276428	BRM CROSSED BRIDGE GOING SOUTH	OBSERVED
7	0700	GL034276428	PREPARED SKETCH OF BRIDGE GL03117631	COMPLETE
8	0900	GL034276428	MISSION COMPLETED -RETURN TO CP	END OF MISSION

DA FORM 5786-R, JUN 89

Figure 4-25. Example of completed DA Form 5786-R.

CHAPTER 5
MISSION PREPARATION

The sniper team uses planning factors to estimate the amount of time, coordinating and effort that must be expended to support the impending mission. Arms, ammunition, and equipment are METT-T dependent.

Section I
PLANNING AND COORDINATION

Planning and coordination are essential procedures that occur during the preparation phase of a mission.

5-1. MISSION ALERT

The sniper team may receive a mission briefing in either written or oral form (FRAGO). Usually, the team mission is stated specifically as to who, what, when, where, and why/how. On receipt of an order, the sniper analyzes his mission to ensure he understands it, then plans the use of available time.

5-2. WARNING ORDER

Normally, the sniper team receives the mission briefing. However, if the sniper receives the briefing, he prepares to issue a warning order immediately after the briefing or as soon as possible. He informs the observer of the situation and mission and gives him specific and general instructions. If the sniper team receives the mission briefing, the sniper should still present the warning order to the observer to clarify and emphasize the details of the mission briefing.

5-3. TENTATIVE PLAN

The sniper makes a tentative plan of how he intends to accomplish the mission. When the mission is complex and time is short, he makes a quick, mental estimate; when time is available, he makes a formal, mental estimate. The sniper learns as much as he can about the enemy and mission requirements and applies it to the terrain in the assigned area. Since an on-the-ground reconnaissance is not tactically feasible for most sniper operations, the sniper uses maps, pictomaps, or aerial photographs of the objective and surrounding area to help formulate his tentative plan. This plan is the basis for team preparation, coordination, movement, and reconnaissance.

5-4. COORDINATION CHECKLISTS

Coordination is continuous throughout the planning phase of the operation (see coordination checklists) (for example, aircraft, parachutes, or helicopters). Other items are left for the sniper to coordinate. He normally conducts coordination at the briefing location. To save time, he assigns tasks to the observer and has him report back with the results. However, the sniper is responsible for all coordination. He uses coordination checklists to verify mission-essential equipment for the mission. He coordinates directly with appropriate staff sections or the S3, or the SEO will provide the necessary information. The sniper may carry a copy of the coordination checklists to ensure he does not overlook an item that may be vital to the mission. Coordination with specific staff sections includes the following:

NOTE: Items may need coordination with more than one staff section; therefore, some items are listed under more than one heading.

a. **Intelligence.** The S2 informs the sniper of any changes in the situation as given in the OPORD or mission briefing. The sniper constantly updates the tentative plan with current information.

(1) Identification of the unit.
(2) Weather and light data.
(3) Terrain update.
   • Aerial photos.
   • Trails and obstacles not on map.
(4) Known or suspected enemy locations.
(5) weapons.
(6) Strength.
(7) Probable courses of action.
(8) Recent enemy activity.
(9) Reaction time of reaction forces.
(10) Civilian activity in area.
(11) Priority intelligence requirements and information requirements.
(12) Challenge and password.

b. **Operations.** The sniper coordinates with the operations section to receive the overall status of the mission.

(1) Identification of the unit.

(2) Changes in the friendly situation.

(3) Route selections and LZ and PZ selections.

(4) Linkup procedure.

(5) Transportation (other than air).

(6) Resupply (along with S4).

(7) Signal plan.

(8) Departure and reentry of forward units.

(9) Special equipment requirements.

(10) Adjacent units operating in the area of operations.

(11) Rehearsal areas.

(12) Method of insertion/extraction.

(13) Frequencies and call signs.

c. **Fire Support.** Usually, the sniper coordinates fire support with the fire support officer.

(1) Identification of the unit.

(2) Mission and objective.

(3) Routes to and from the objective (including alternate routes).

(4) Time of departure and expected time of return.

(5) Unit target list (fire plan).

(6) Fire support means available (artillery, mortar, naval gunfire, and aerial fire support to include Army, Navy, and Air Force).

(7) Ammunition available (to include different fuzes).

(8) Priority of fires.

(9) Control measures for fire support.

- Checkpoints.
- Boundaries.
- Phase lines.
- Fire support coordination measures.
- Priority targets (list TRPs).
- RFA.
- RFL.
- No-fire areas.
- Precoordinate authentication.

(10) Communications (include primary and alternate means, emergency signals, and code words and signals).

d. **Coordination with Forward Unit.** A sniper team that must move through a friendly forward unit must coordinate with the unit commander for a smooth, orderly passage. If there is no coordination time and place, the sniper sets the time and place with the S2 and S3. Then, he informs the forward unit and arranges assistance for the team's departure. Coordination is a two-way exchange of information.

(1) Identification (team leader, observer, and unit).

(2) Size of team.

(3) Time(s) and place(s) of departure and return, location(s) of departure point(s), IRPs, and detrucking points.

(4) General area of operation.

(5) Information on terrain and vegetation.

(6) Known or suspected enemy positions or obstacles.

(7) Possible enemy ambush sites.

(8) Latest enemy activity.

(9) Detailed information on friendly positions (for example, crew-served weapons or final protective fire).

(10) Fire and barrier plan.

(11) Support the forward unit can furnish. How long and what can they do?

- Fire support.
- Litter teams.
- Navigational signals and aids.
- Guides.
- Communications.
- Reaction units.
- Other.

(12) Call signs and frequencies and exchange of Vinson cryptographic variables.

- Pyrotechnic plans.
- Challenge and password.
- Emergency signals and codewords.
- Relieved unit (pass information to the relieving unit).

e. **Adjacent Unit Coordination.** Immediately after receiving the OPORD or mission briefing, the sniper coordinates with other units using the same area. If he is not aware of other units, he should check with the S3 to arrange coordination. The sniper exchanges the following information with other units or snipers operating in the same area:

(1) Identification of the unit.

(2) Mission and size of unit.

(3) Planned times and points of departure and reentry.

(4) Route.

(5) Fire support (planned) and control measures.

(6) Frequency, call signs, and exchange of Vinson cryptographic variables.

(7) Challenge and password and or number.

(8) Pyrotechnic plans.

(9) Any information that the unit may have about the enemy.

f. **Rehearsal Area Coordination.** The sniper coordinates with the S2 or S3.

(1) Identification of own unit.

(2) Mission.

(3) Terrain similar to objective site.

(4) Security of the area.

(5) Availability of aggressors.

(6) Use of blanks, pyrotechnics, live ammunition.

(7) Mockups available.

(8) Time the area is available (preferably when light conditions are close to the expected light conditions for the mission).

(9) Transportation.

(10) Coordination with other units using the area.

g. **Army Aviation Coordination.** The sniper coordinates with the supporting aviation unit commander through the S3 or S3 Air.

(1) *Situation:*

(a) *Enemy forces:* Location, activity, probable course of action, and enemy air defense.

(b) *Weather:* Decision time/ POC any delay for the mission.

(c) *Friendly forces:* Main mission, activity, boundaries, axis of movement.

(2) *Mission:* Task and purpose.

(3) *Execution:*

(a) *Concept of the operation:* Overview of what requesting unit wants to accomplish with the air assault/ air movement.

(b) *Coordinating instructions (PZ operation):*
- Direction of landing.
- Time of landing/ flight direction.
- Location of PZ/ alternate PZ.
- Loading procedures.
- Marking of PZ (panel, smoke, smoke munitions, lights).

- Flight route planned (start point, air control point, rally point).
- Formation of landing/ flight/ landing (LZ).
- Code words: PZ secure (before landing); PZ clear (lead plane, last plane); alternate PZ (at PZ en route, at landing zone); names of PZ/ alternate PZ.
- TAC air/ artillery.
- Number of passengers/ planes for entire lift.
- Equipment carried by individuals.
- Secure PZ or not.
- Marking of key leaders (LZ operations).
- Direction of landing.
- Time of landing, false insertions.
- Location of LZ or alternate LZ.
- Marking of LZ (panel, smoke, SM, lights).
- Formation of landing.
- Codewords: LZ name, alternate LZ name.
- TAC air/ artillery preparation, fire support coordination.
- Secure LZ or not.

(4) *Service support:*
(a) Number of aircraft, times, number of lifts.
(b) Refuel/ rearm during mission or not.
(c) Special equipment/ aircraft configuration for weapons earned by unit personnel.
(d) Bump plan.

(5) *Command and signal:*
(a) Frequency and call signs.
(b) Location of air mission commander.

h. **Vehicle Movement Coordination.** The sniper coordinates with the supporting unit through the S3.
(1) Identification of the unit.
(2) Supporting unit identification.
(3) Number and type of vehicles and tactical preparation.
(4) Entrucking point.
(5) Departure/ loading time.
(6) Preparation of vehicles for movement.
- Driver responsibilities.
- Sniper team responsibilities.
- Special supplies/ equipment required.

(7) Availability of vehicles for preparation/ rehearsal/ inspection (time and location).

(8) Route
- Primary.
- Alternate.
- Checkpoints.

(9) Detrucking points.
- Primary.
- Alternate.

(10) March interval/ speed.

(11) Communications (frequencies, call signs, codes).

(12) Emergency procedures and signals.

5-5. COMPLETION OF PLAN

After the warning order has been issued and a thorough map reconnaissance made, most coordination should be completed. The sniper makes an intelligence update while the observer prepares himself and the equipment for the mission. The sniper completes his plan based on his map reconnaissance and or any changes in the enemy situation. He may or may not alter the tentative plan, but he can add detail. The sniper uses the OPORD format as a guide to refine his concept. He places the main focus on actions in the objective area and carefully assigns the observer specific tasks for all phases of the operation. He ensures all actions work smoothly and efficiently.

5-6. OPERATION ORDER

The operation order is issued in the standard OPORD format. Extensive use of terrain models, sketches, and chalkboards should be made to highlight important details such as routes, planned rally points, and actions at known danger areas. All aspects of the OPORD should be thoroughly understood by the sniper team to include memorizing the following

- Intelligence acquisition tasks.
- Situation—both friendly and enemy.
- Mission.
- Execution plan.
- Administrative plans.
- Communications and electronics, including frequencies, call signs, and antennas to be used.
- SALUTE format.

5-7. BRIEFBACK

The sniper team rehearses the briefback until it is near-perfect before presenting it to the S3, sniper employment officer, or commander. A good briefback indicates the team's readiness for the mission. (Figure 5-1 is an example of a sniper team briefback outline.)

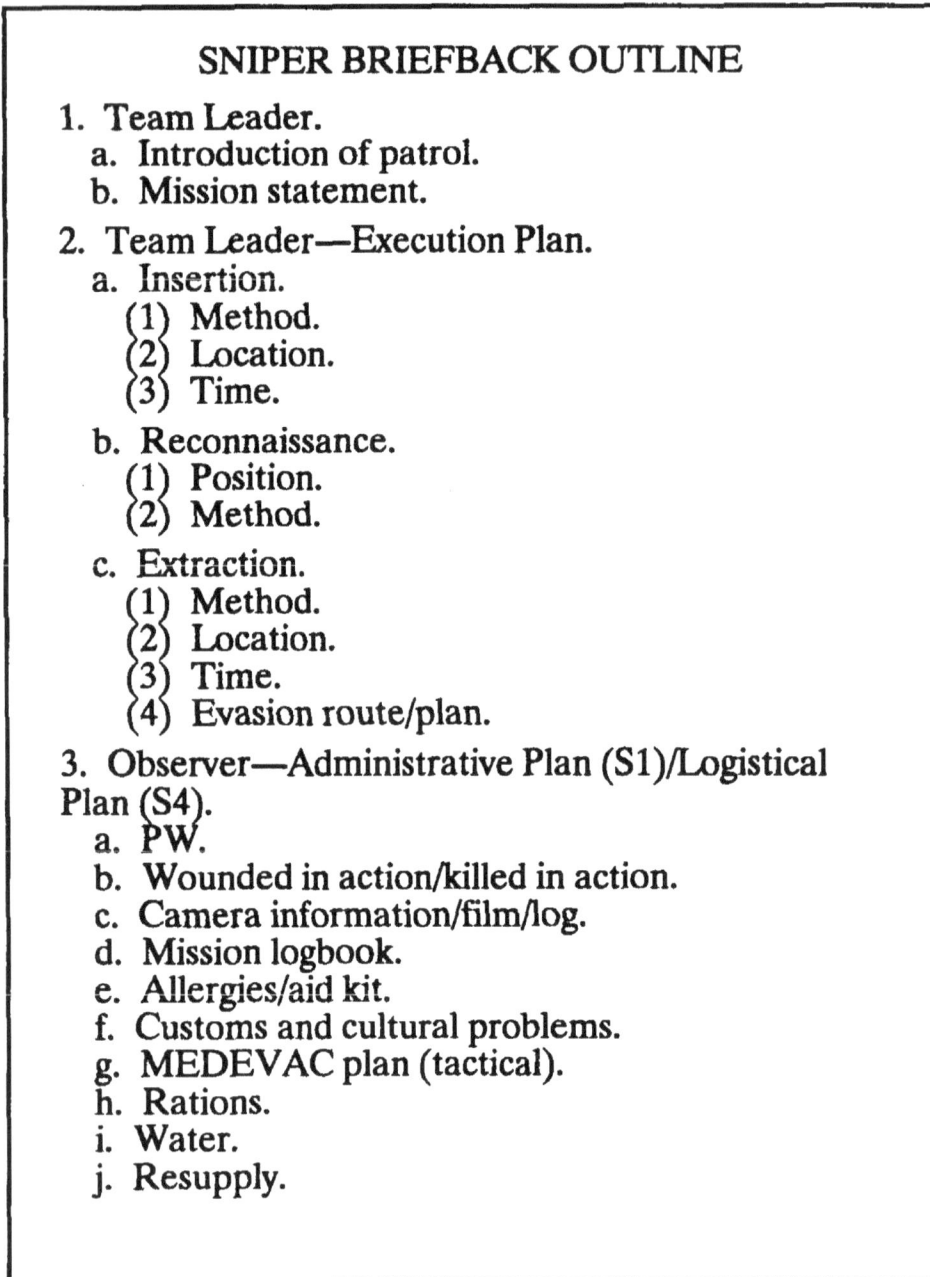

SNIPER BRIEFBACK OUTLINE

1. Team Leader.
   a. Introduction of patrol.
   b. Mission statement.
2. Team Leader—Execution Plan.
   a. Insertion.
      (1) Method.
      (2) Location.
      (3) Time.
   b. Reconnaissance.
      (1) Position.
      (2) Method.
   c. Extraction.
      (1) Method.
      (2) Location.
      (3) Time.
      (4) Evasion route/plan.
3. Observer—Administrative Plan (S1)/Logistical Plan (S4).
   a. PW.
   b. Wounded in action/killed in action.
   c. Camera information/film/log.
   d. Mission logbook.
   e. Allergies/aid kit.
   f. Customs and cultural problems.
   g. MEDEVAC plan (tactical).
   h. Rations.
   i. Water.
   j. Resupply.

Figure 5-1. Briefback outline.

4. Observer—Intelligence (S2).
   a. Enemy situation.
   b. Friendly situation.
   c. Weather.
   d. Terrain restrictions (including MEDEVAC restrictions).
   e. Light data, percentage of illumination.
   f. Order of battle.
   g. Intelligence acquisition task.

5. Observer—Communications and Electronics.
   a. Window times and format.
   b. Frequencies and call signs.
   c. Antenna-type and terrain considerations.
   d. Special equipment.
   e. Code words.

6. Team Leader—Conclusion.
   a. Restate mission.
   b. Combat medical plan.
   c. Special Instructions Relating to Training Mission.
      (1) Contact with civilians.
      (2) Local authority POC.
      (3) Maneuver restrictions.

Figure 5-1. Briefback outline (continued).

5-8. EQUIPMENT CHECK

The sniper team ensures needed equipment is operational before signing it out. Weapons are clean, functional, and test-fired to confirm zero. The team checks radios by making a communications check with the NCS of the net they will be using and night vision devices by turning them on (adding an extra battery). The team then double-checks all equipment. If they encounter problems, the sniper notifies the PSG or SEO.

5-9. FINAL INSPECTION

Inspections reveal the team's physical and mental state of readiness. The sniper ensures that all required equipment is present and functional, and that the observer knows and understands the mission. The following items should be inspected:

- Completeness and correctness of uniform and equipment.
- Items such as pictures, papers, marked maps, and sniper data book that contain confidential material.
- Hats and pockets.
- Shine, rattles, and tie-downs.
- Weapons (loaded or unloaded).
- Fullness of canteens.

If unauthorized items are found, the sniper immediately corrects any deficiencies. Then, he questions the observer to make sure he knows the team plan, what his job is, and when he is to do it.

5-10. REHEARSALS

Rehearsals ensure team proficiency. During rehearsals, the sniper rechecks his plans and makes any needed changes. It is through well-directed and realistic rehearsals that the team becomes thoroughly familiar with their actions on the mission.

a. The sniper team uses terrain similar to that on which they will operate (if available), rehearsing all actions if time permits. A good way to rehearse is to talk the team through each phase, describing the actions of each sniper, and then perform the actions as a dry run. When actions are understood, the sniper team goes through all the phases, using the signals and commands to be used during the mission.

b. If there is no time for rehearsals, the sniper team conducts a briefback/ talk-through. This method is used to supplement rehearsals or when security needs or a lack of time preclude dry runs and wet runs. In this method, the *team leader* talks the *observer* through his actions and then has him orally repeat those actions. The sniper team establishes the sequence of actions to be rehearsed and, if time permits, conducts rehearsals in the same sequence as in the mission.

5-11. FINAL PREPARATIONS

The sniper makes any last-minute changes and corrects any deficiencies found during initial inspections. Final inspection should be made by the SEO and an S3 representative. Again, pockets and rucksacks are emptied and inspected. The inspection team looks for personal papers, marked maps, and other unauthorized items. The sniper ensures all previous discrepancies are corrected; equipment is still operational; all needed items are present; and the observer is ready for the mission. The inspection team randomly asks questions about the mission.

5-12. PREPARATION FOR DEBRIEFING

After the mission, the SEO or S3 representative directs the sniper team to an area where they prepare for a debriefing. The team remains in the area until called to the operations center. The sniper will bring the sniper data book that contains a log sheet, a field sketch, a range card, and a road/area sketch for debriefing.

a. The sniper team—

(1) Lays out and accounts for all team and individual equipment.

(2) Consolidates all captured material and equipment.

(3) Reviews and discusses the events listed in the mission logbook from insertion to return, including details of each enemy sighting.

(4) Prepares an overlay of the team's route, area of operations, insertion point, extraction point, and significant sighting locations.

b. An S3 representative controls the debriefing. He directs the sniper—

(1) To discuss any enemy sightings since the last communications with the radio base station.

(2) To give a step-by-step amount of each event listed in the mission logbook from insertion until reentry of the FFL, including details of all enemy sightings.

c. When the debriefing is complete, the S3 representative releases the sniper team back to platoon control.

5-13. COUNTERSNIPER OPERATION

When an enemy sniper threat has been identified in the sniper team's area of operations, the team is employed to eliminate the enemy sniper.

a. A sniper team identifies an existing sniper threat by using the following indicators:

(1) Enemy soldiers in special camouflage uniforms.

(2) Enemy soldiers seen carrying weapons in cases or drag bags or weapons with long barrel lengths, mounted telescopes, and bolt-action receivers.

(3) Single-shot fire.

(4) Lack or reduction of enemy patrols during single-shot fire.

(5) Light reflecting from optical lenses.

(6) Reconnaissance patrols reporting small groups of (one to three) enemy soldiers.

(7) Discovery of single expended casings, such as 7.62-mm ammunition.

b. The sniper team then determines the best method to eliminate the enemy sniper. To accomplish this, the team gathers information and determines the pattern.

(1) *Gathers information.*

(a) Time of day precision fire occurs.

(b) Location of encountered enemy sniper fire.

(c) Location of enemy sniper sightings.

(d) Material evidence of enemy snipers, such as empty brass casings or equipment.

(2) *Determines patterns.* The sniper team evaluates the information to detect established patterns or routines. The team conducts a map reconnaissance, studies aerial photographs, or carries out ground reconnaissance to determine the movement patterns. The sniper must place himself in the position of the enemy and ask, "How would I accomplish this mission?"

c. Once a pattern or routine is identified, the sniper team determines the best location and time to engage the enemy sniper. The team can also request the following:

(1) Coordinating routes and fires.

(2) Additional preplotted targets (fire support).

(3) Infantry support to canalize or ambush the enemy sniper.

(4) Additional sniper teams for mutual supporting fire.

(5) Baiting of likely engagement areas to deceive the enemy sniper into commitment by firing.

(6) All elements in place 12 hours before the expected engagement time.

During a countersniper operation, the team must ignore battle activity and concentrate on the enemy sniper.

d. When an enemy sniper is operating in a unit's area, the sniper team ensures the unit employs passive countermeasures to defend against enemy sniper fire.

(1) Do not establish routines. For example, consistent meal times, ammunition resupply, assembly area procedures, or day-to-day activities that have developed into a routine.

(2) Conduct all meetings, briefings, or gatherings of personnel undercover or during limited visibility.

(3) Cover or conceal equipment.

(4) Remove rank from helmets and collars. Do not salute officers. Leaders should not use authoritative methods.

(5) Increase OPs and use other methods to increase the unit's observation abilities.

(6) Brief patrols on what to look for, such as single, expended rounds or different camouflage materials.

(7) Do not display awareness of the enemy's presence at any time.

5-14. REACTION TO ENEMY SNIPER FIRE

Although the sniper team's mission is to eliminate the enemy sniper, the team avoids engaging in a sustained battle with the enemy sniper. If the team is pinned down by enemy sniper fire and the sniper's position cannot be determined, the sniper team attempts to break contact to vacate the enemy sniper's kill zone.

a. The sniper team uses either hand-held or artillery generated smoke to obscure the enemy sniper's view. If the smoke provides sufficient obscuration, the sniper team breaks contact and calls for indirect fire on the enemy sniper position. If the smoke does not provide sufficient obscuration, the sniper team calls for an immediate suppression mission against the enemy sniper position. The team then breaks contact under the cover of indirect fire.

b. The sniper team should expect indirect fire and increased enemy patrolling activity shortly after contact with an enemy sniper.

Section II
MISSION PACKING LISTS

The sniper team requires arms and ammunition as determined by METT-T Some of the equipment mentioned in the example lists may not be available. A sniper team carries only mission-essential equipment normally not associated with a standard infantryman.

5-15. ARMS AND AMMUNITION

As a minimum, the sniper team requires arms and ammunition that should include the following

a. **Sniper.**
- M24 sniper weapon system with M3A scope.
- M9 bayonet.
- 100 rounds M118 special ball.
- M9 pistol.

- 45 rounds 9-mm ball ammunition.
- 4 M67 fragmentation grenades; 2 CS grenades; 2 concussion grenades (MOUT).
- M18A1 mine, complete.

b. **Observer:**
- M16A1/ A2/ M203 with quadrant sight and AN/ PVS-4 mounted.
- M9 9-mm pistol.
- M9 bayonet.
- 210 (plus) rounds 5.56-mm ball ammunition.
- 45 rounds 9-mm ball ammunition.
- 6 rounds 40-mm high-explosive ammunition.
- 3 rounds 40-mm antipersonnel ammunition.
- 4 M67 fragmentation grenade, 2 CS grenades; 2 concussion grenades (MOUT).

5-16. SPECIAL EQUIPMENT
The sniper team requires special equipment that may include, but not be limited to the following

a. **Sniper:**
- M24 sniper weapon system deployment kit (tools and replacement parts).
- M9 pistol cleaning kit.
- Extra handset for radio.
- Extra batteries for radio (BA 4386 or lithium, dependent on mission length).
- SOI.
- M15 tripod.
- M49 observation telescope.
- AN/ PVS-5/ 7 series, night vision goggles.
- Extra BA-1567/ U or AA batteries for night vision goggles.
- Pace cord.
- E-tool with carrier.
- 50-foot 550 cord.
- 1 green and 1 red star cluster.
- 2 HC smoke grenades.
- Measuring tape (25-foot carpenter-type).
- 3 each 9-mm magazines

b. **Observer:**
- M16A1/ A2 cleaning kit.
- M203 cleaning kit.
- AN/ PRC-77/ AN-PRC-119/ AN/ PRC-104A radios.
- Radio accessory bag, complete with long whip and base, tape antenna and base, handset, and battery (BA-4386 or lithium).
- 300-feet WD-1 field wire (for field-expedient antenna fabrication).
- Olive-drab duct tape ("100-mph" tape).
- Extra batteries for radio (if needed).
- Extra batteries (BA-1567/ U) for AN/ PVS-4.
- M19/ M22 binoculars.
- Sniper's data book, mission logbook, range cards, wind tables, and "slope dope."
- 7 each 30-round capacity (5.56-mm) magazines.
- 3 each 9-mm magazines.
- Calculator with extra battery.
- Butt pack.
- 10 each sandwich-size waterproof bags.
- 2 HC smoke grenades.
- Lineman's tool.
  Range estimation (sniper data book).

5-17. UNIFORMS AND EQUIPMENT
A recommended listing of common uniforms and equipment follows; however, weather and terrain will dictate the uniform. As a minimum, the sniper team should have the following
- Footgear (jungle/ desert/ cold weather/ combat boots).
- 2 sets BDUs (desert/ woodland/ camouflage).
- Black leather gloves.
- 2 brown T-shirts.
- 2 brown underwear (optional).
- 8 pair olive-drab wool socks.
- Black belt.
- Headgear (BDU/ jungle/ desert/ cold weather).
- ID tags and ID card.
- Wristwatch (sweep second hand with luminous dial/ waterproof).
- Pocket survival knife.

- Extra large ALICE pack, complete with frame and shoulder straps.
- 2 waterproof bags (for ALICE pack).
- 2 two-quart canteens with covers.
- 1 bottle water purification tablets.
- LBE complete.
- Red-lensed flashlight (angle-head type with extra batteries).
- MREs (number dependent on mission length).
- 9-mm pistol holster and magazine pouch (attached to LBE).
- 2 camouflage sticks (METT-T dependent).
- 2 black ink pens.
- 2 mechanical pencils with lead.
- 2 black grease pencils.
- Lensatic compass.
- Map(s) of operational area and protractor.
- Poncho.
- Poncho liner.
- 1 each ghillie suit complete.
- 1 each protective mask/ MOPP suit.
- Foot powder.
- Toiletries.
- FM 23-10.

5-18. OPTIONAL EQUIPMENT

Certain situations may require equipment for specialized tasks and is METT-T dependent. The following equipment may prove useful in different climates/ operational areas:

- M203 vest.
- Desert camouflage netting.
- Natural-colored burlap.
- Glitter tape.
- VS-17 panel.
- Strobe light with filters.
- Special patrol insertion/ extraction system harness.
- 12-foot sling rope.
- 2 each snap links.
- 120-foot nylon rope.

- Lip/ sunscreen.
- Signal mirror.
- Pen gun with flares.
- Chemical lights (to include infrared).
- Body armor/ flak jacket.
- Sniper veil.
- Sewing kit.
- Insect repellant.
- Sleeping bag.
- Knee and elbow pads.
- Survival kit.
- Rifle drag bag.
- Pistol silencer/ suppressor.
- 2.5-pounds C4 with caps, cord, fuze, and igniter.
- Rifle biped/ tripod.
- Empty sandbags.
- Hearing protection (earmuffs).
- Thermometer.
- Laser range finder.
- Thermal imager.
- Pocket binoculars.
- 35-mm automatic loading camera with appropriate lenses and film.
- ½-inch camcorder with accessories.
- Satellite communications equipment.
- Short-range radio with earphone and whisper microphone.
- Field-expedient antennas.
- Information reporting formats.
- Encryption device for radio.

5-19. SPECIAL TOOLS AND EQUIPMENT (MOUT)

For operations in urban areas, the following tools and equipment are most useful; however, they are subject to availability

- Pry bar.
- Pliers.
- Screwdriver.
- Rubber-headed hammer.

- Glass cutter.
- Masonry drill and bits.
- Metal shears.
- Chisel.
- Auger.
- Lock pick, skeleton keys, cobra pick.
- Bolt cutters.
- Hacksaw or handsaw.
- Sledgehammer.
- Axe.
- Ram.
- Power saw.
- Cutting torch.
- Shotgun.
- Spray paint.
- Stethoscope.
- Maps/ street plans.
- Photographs, aerial and panoramic.
- Whistle.
- Luminous tape.
- Flex cuffs.
- Padlocks.
- Intrusion detection system (booby traps).
- Portable spotlights.
- Money.
- Civilian attire.

5-20. ADDITIONAL EQUIPMENT TRANSPORT

The planned use of air and vehicle drops and caching techniques eliminates the need for the sniper team to carry extra equipment. Another method is to use the stay-behind technique when operating with a security patrol. (See Chapter 7.) Through coordination with the security patrol leader, the team's equipment may be distributed among the patrol members. On arrival at the ORP, the security patrol may leave behind all mission-essential equipment. After completing the mission, the team may cache the equipment for later pickup, or it may be returned the same way it was brought in.

CHAPTER 6
OPERATIONS

The SEO aids the sniper team in coordination of air support available for the three phases of operations: insertion, execution, and extraction and recovery. These techniques may be limited by the type of unit to which the sniper team is assigned, depending on the unit's resources. The team should adhere to the plan outlined in this chapter.

Section I
INSERTION

Insertion is the first critical phase of sniper operations. Regardless of the mission, the team must pass through terrain where the enemy may use sophisticated detection devices. The selected method of insertion depends on the mission, enemy situation, resources available, weather and terrain, depth of penetration, and mission priority.

6-1. PLANNING INSERTION

The preferred method of insertion is the one that best reduces the chance of detection. To provide the most current and specific details on the target area and infiltration routes from all sources, the headquarters and the sniper team adhere to the following:

a. **Intelligence.** Base operational plans on timely and accurate intelligence. Place special emphasis on efforts to obtain information on the enemy's ability to detect forces inserted by air, water, or land. The location and capabilities of air defense radar and weapons systems are critical.

b. **Deception.** Make plans to deny the enemy knowledge of the sniper team's insertion or to deceive him as to the location or intent of the operation. False insertions and other cover operations (such as air strikes, ground attacks, and air assault operations), as well as the use of multiple routes and means of insertion, ECM, and false transmissions, contribute to sniper deception plans. Select unexpected means of insertion, times, places, and routes, coupled with speed and mobility to help deceive the enemy. Also include in plans diversionary fires to direct the enemy's attention away from the team. Specific techniques may include the following

(1) Multiple airdrops, water landings, or both.

(2) Dispersion of insertion craft (air or water) if more than one, both in time and location.

(3) Landing a force in an area closer to other potential targets than to the actual targets.

(4) Leaks of false information.

(5) False landings or insertions.

(6) Diversionary actions, such as air strikes in other areas.

(7) Increased reconnaissance flights over false areas.

c. **Speed and Mobility.** Tailor individual loads to enhance speed and mobility, and balance these loads with the mission-related items necessary to achieve success. Speed is essential to limit the amount of time required to insert the team. If possible, carry only what is needed immediately and cache the rest to be retrieved.

d. **Stealth.** Stress stealth to avoid detection or interception by the enemy at the time of insertion during movement along routes and while traveling from the insertion area to the target area.

e. **Suppression.** Suppress enemy detection devices, weapons systems, and command and control facilities by electronic jamming or by suppressive fires. This detracts from the enemy's ability to discover the team during infiltration. Deception techniques contribute to suppression activities.

f. **Security.** Emphasize security measures to prevent compromise of the impending operation during preparation. This includes the security of rehearsal and training sites. Some measures that maybe used to assist in maintaining security areas follows:

(1) Restrict access to the isolation area during planning.

(2) Brief details of the operation to the team in the isolation area.

(3) Limit knowledge of planned operations on a need-to-know basis.

g. **Reconnaissance, Surveillance, and Target Acquisition.** Increase the use of RSTA equipment to detect and avoid enemy forces and their detection devices. Use passive night vision devices to achieve rapid assembly and reorganization. Also use these devices to help control and speed of movement and to traverse seemingly impassable terrain.

h. **Rehearsals.** Ensure rehearsals parallel, as close as possible, the actual conditions of insertion or extraction. Conduct rehearsals on terrain similar to that in the target area.

i. **Sand Tables.** Use sand tables in the planning phase since they are effective for orienting the team on unfamiliar DZs and surrounding terrain. The use of sand tables and terrain models enhance orderly and rapid assembly on the ground during the issuance of prejump orders and briefings.

6-2. AIR INSERTION

Air insertion is the fastest way to infiltrate. Sniper teams and equipment may be delivered by parachute (static-line or free-fall technique), fixed-wing (air landing), or helicopter (air landing, rappelling, or parachuting).

a. **Special Factors.** When planning an air insertion, headquarters considers several factors.

(1) A primary danger area is the perimeter (frontier area) where the enemy uses the most sophisticated weapons systems and air defenses.

(a) Suppression of enemy air defense maybe necessary along the infiltration corridor. This is done by a variety of sophisticated counter-measures applied against enemy equipment and by strikes against known enemy positions. Artillery, aircraft, or naval gunfire may provide assistance.

(b) Fire support, smoke screens, and suppressive measures may be critical since most of the enemy's detection devices and air defense weapons may be near the point of entry. Special equipment may be required to counter the enemy's RSTA effort whether moving by air, water, or land.

(2) If this area is within artillery or NGF range, fires should be planned on known and suspected enemy antiaircraft locations and on prominent landforms along the route.

(3) All flights over enemy territory should be routed over unoccupied areas, if possible. Flights should be planned to complement cover and deception phases and to avoid enemy air defenses.

(4) Since the sniper team depends on the transporting unit during this phase, snipers must coordinate all aspects of the air insertion with these units. To lower the chances of detection, the team makes the greatest use of reduced visibility, tactical cover, and deception. Drop zones and landing zones should be behind tree lines, in small forest clearings, or on other inconspicuous terrain.

(5) The sniper team considers the chance of in-flight emergencies. It must know the route and the checkpoints along the route. The team establishes simple ground assembly plans for contingencies before boarding. In an emergency, the SEO decides whether to continue or

abort the mission. In the absence of the SEO, the sniper makes the decision based on METT-T factors, contingency plans, and the distance to the target as compared to the distance back to forward friendly lines. Contingency provisions should be made for air and water rescue as well.

b. **Special Airborne Assault Techniques.** In airborne insertions during limited visibility, the headquarters emphasizes special delivery or navigational techniques.

(1) With the AWADS, personnel and equipment can be air-dropped during bad weather, even during zero-visibility conditions. Insertions may be made (day or night) without a pre-positioned USAF combat control team or an Army assault team. The supporting air unit requires both extensive DZ intelligence and significant lead time. All forces involved must thoroughly plan and coordinate the operation.

(2) HALO or HAHO jumps with high-performance parachutes allow parachutists to maneuver to a specific point on the ground. During these operations, they can use midair assembly procedures.

c. **Assembly.** The sniper team must be able to assemble and reorganize quickly and precisely because of its vulnerability to detection. The team develops assembly plans after careful consideration of METT-T factors, especially the location of the enemy, visibility, terrain, DZ information, dispersion pattern, and cross-loading. The number of assembly areas depends on the location, the size of available assembly areas, and the enemy's detection ability.

(1) Terrain association may be used as a backup method of designating assembly areas, but it has obvious disadvantages if the unit misses the DZ or if an in-flight change in mission dictates use of a new drop zone.

(2) A night vision plan is needed during landing, assembly, and movement in reduced visibility.

(3) Cold weather airborne insertion is difficult. Allocated times must be increased by at least 30 minutes for cold weather insertions.

(4) The team must be aware of the location of the assembly areas in relation to the direction of flight of the insertion aircraft. The direction of flight is 12 o'clock.

(5) During parachute insertion, team members must be ready for enemy engagement at all times, especially on the DZ. Immediate-action drills are required to counter enemy contact on the DZ.

d. **Planning.** The reverse planning process is of paramount importance for the ground tactical plan. The ground tactical plan, as developed from the mission assessment, is the first planning area to be considered. All other planning begins from this point.

(1) The selection of PZs or LZs requires adequate planning and coordination for effective use of air assets. Site selection must be coordinated face-to-face between the sniper team and the supporting aviation commander. The tactical situation is the key planning factor; others include the following:

- Size of landing points.
- Surface conditions.
- Ground slopes
- Approach and departure directions.
- Aircraft command and control.
- PZ and LZ identification.
- Rehearsals.

(2) The air movement plan coordinates movement of the team into the zone of action in a sequence that supports the landing plan. Key considerations are flight routes, air movement tables, flight formation, in-flight abort plan, altitude, and air speed.

(3) The landing plan introduces the team into the target area at the proper time and place. Rehearsals cannot be overemphasized. The team rapidly assembles, reorganizes, and leaves the insertion site. Fire support, if available, may be artillery, NGF, attack helicopters, or USAF tactical aircraft. The fire support plan must support all other plans. Supporting fires must be thoroughly coordinated with the air mission commander. Other planning considerations are evasion and escape, actions at the last LZ, assembly plan, downed aircraft procedures, control measures, weather delays, deception plans, and OPSEC.

6-3. AMPHIBIOUS INSERTION

Water insertion may be by surface swimming, small boat, submarine, surface craft, helocasting, or a combination thereof. The sniper team needs detailed information to plan and execute a small-boat landing, which is the most difficult phase of a waterborne insertion. Close coordination is required with naval support units.

a. **Planning.** While on the transporting craft, the team plans for all possible enemy actions and weather. Initial planning includes the following:

(1) *Time schedule.* The time schedule of all events from the beginning until the end of the operation is used as a planning guide. Accurate timing for each event is critical to the success of the operation.

(2) *Embarkation point.* The embarkation point is the point where the team enters the transporting craft.

(3) *Drop site.* The drop site is the site where the team leaves the primary craft and loads into a smaller boat.

(4) *Landing site.* The landing site is the site where the team beaches the boat or lands directly from amphibious craft.

(5) *Loading.* Loads and lashings, with emphasis on waterproofing, are followed IAW unit SOPs. Supervisors must perform inspections.

b. **Beach Landing Site Selection.** The beach landing site must allow undetected approach. When possible, the team avoids landing sites that cannot be approached from several different directions. The site chosen allows insertion without enemy detection. If sand beaches are used, tracks and other signs must be erased that may compromise the mission. Rural, isolated areas are preferred. The coastal area behind the landing site should provide a concealed avenue of exit. Other factors considered in each selection include enemy dispositions, distance to the target area, characteristics of landing and exit sites, and availability of cover and concealment.

c. **Tactical Deception.** Besides the water approach route plan, plans must deny the enemy knowledge of the insertion. This may include use of ECM or diversionary fire support to direct the enemy's attention away from the insertion site.

d. **Routes.** The route to the drop site should be planned to deceive the enemy. If possible, the route should be similar to that used in other types of naval operations (minelaying, sweeping, or patrolling). A major route change immediately after the team's debarkation may compromise the mission.

e. **Navigation.** Ship-to-shore navigation (to the landing site) maybe accomplished by dead reckoning to a shoreline silhouette or radar.

f. **Actions at the Drop Site.** Primary and alternate drop sites must be agreed upon. The drop site should be at least 1,500 meters offshore to prevent compromise by noise during loading and launching. (Some operations may permit landing directly from the transporting craft on shore.) If the enemy has surface radar capability, the drop site may need to be several miles offshore, or the use of ECM may be required.

g. **Actions at the Beach Landing Site.** To plan actions at the landing site, the team must consider the following:

- Actions during movement to the beach.
- Noise and light discipline.
- Navigational techniques and responsibilities.
- Actions on the beach.
- Plan for unloading boats (SOP).
- Plan for disposal or camouflage of boats.

h. **Actions on the Beach.** Once on the beach, sniper team members move to a covered and concealed security position to defend the landing site. The sniper team then conducts a brief listening halt and checks the beach landing area for signs of enemy activity. The team may deflate, bury, or camouflage the boat near the landing site or away from it, depending on the enemy situation, the terrain, and the time available. If the boat is to be disposed of or hidden near the landing site, a member must be designated to dig a hole or cut brush for camouflage. After the boat is disposed of, a designated team member sweeps the beach to erase tracks and drag marks.

i. **Insertion by Air From Ship.** Helicopters launched from a ship may extend the range of sniper teams. They may be vectored from ships to a predetermined LZ. Once in the air, other aspects of landing and assembling are the same as for air movement operations.

j. **Helocasting.** Helocasting combines a helicopter and small boat in the same operation. It is planned and conducted much the same as air movement operations, except that the LZ is in the water. While a helicopter moves at low levels (20 feet) and low speeds (20 knots), the sniper team launches a small boat and enters the water. Members then assemble, climb into the boat, and continue the mission.

k. **Contingency Planning.** The following contingencies must be covered in the planning stage:

- Enemy contact en route.
- Hot helocast site.
- Flares.
- Aerial attack.
- Small arms.
- Indirect fire.
- Downed aircraft procedure (if applicable).
- Evasion and escape.
- High surf.
- Adverse weather.
- Separation.

l. **Rehearsals.** The team must rehearse all aspects of the amphibious insertion to include boat launching, paddling, boat commands, capsize drills, beaching, and assembly.

6-4. LAND INSERTION

Land insertion from a departure point to the target area sometimes may be the best (or only) way to accomplish a mission. Normally, this is so when the enemy has total air superiority or has established effective air defenses. The sniper team can accomplish land insertions over any type of terrain, in any climate. However, thick forests, swamps, and broken or steep terrain probably offer the best chance of success.

a. **Planning.** Plans for overland movement enable the sniper team to move to the target area with the least risk of detection. Planning considerations include the following

(1) Selecting concealed primary and alternate routes based on detailed map reconnaissance and aerial photographs, ground reconnaissance, and data on the enemy situation from other sources.

(2) Avoiding obstacles, populated areas, silhouetting enemy positions, main avenues of approach, and movements along heavily populated routes and trails.

(3) Selecting the time of insertion to take advantage of reduced visibility and reduced alertness. The time is especially important during critical phases while passing through populated areas.

(4) Knowing routes, rendezvous points (and alternates), time schedules, danger areas, and the enemy situation are critical to speed and stealth.

(5) Providing centralized coordination to ensure that members act IAW cover and deception plans. Insertion by land is characterized by centralized planning and decentralized execution.

b. **Actions on Enemy Contact.** Once beyond the FFL, the sniper team must be alert to avoid detection while en route to the target area. If the sniper team becomes aware of the enemy, it must try to move away without an alert. The sniper team fights only when there is no alternative. Then, it breaks contact as quickly as possible. Following enemy contact, the sniper contacts the SEO for a decision to abort or continue the mission. If continuing the mission, the sniper team may have to establish a temporary position for resupply, extraction, or evacuation of wounded.

c. **Stay-Behind Technique.** The sniper team applies the stay-behind technique when the team moves with a security patrol. The team establishes an ORP, caches nonessential equipment, and changes into ghillie suits to prepare for movement to the TFFP. Once this is accomplished, the security patrol departs for a predetermined location to act as a quick-reaction force for the team or returns to its operational base. Use of this technique requires the following considerations:

- Noise and light discipline.
- Avoidance of enemy contact.

- Timing.
- Rough, inaccessible terrain.
- Medical evacuation.
- Communications.
- Method of extraction.
- Evasion and escape.

d. **Actions at the Insertion.** The sniper team develops a detailed assembly plan, basing it on the insertion method and the terrain at the insertion site.

(1) The sniper team selects an assembly area that can be identified at night and is near the insertion site. It uses this assembly area if team members become separated during the insertion. During parachute insertion, the sniper team uses the assembly area as an assembly point.

(2) The sniper team also designates an initial rally point that can be identified at night. The rally point is normally no closer than several hundred meters from the insertion site. The team uses the IRP for assembly if the insertion site is attacked either on insertion or shortly after departing the insertion site.

(3) When the insertion is complete, the sniper team accounts for equipment and supplies, and ensures any injuries are treated. If a disabling injury occurs during insertion, the sniper must decide, based on guidance, whether to continue the mission or to request extraction.

(4) The sniper team's most critical task is verifying the team's location. The sniper verifies his location at the insertion site or after moving away from the site.

(5) The sniper team sterilizes the site and caches or discards nonessential equipment. The preferred method is to bury discards away from the insertion site. The sniper team must camouflage the cache site.

(6) The sniper team departs the insertion site, then halts to listen for sounds of pursuit and to become familiar with local sounds. It establishes a primary azimuth and immediately begins information collection activities and map update.

6-5. VEHICLE INSERTION

Vehicle insertion uses wheeled or tracked vehicles to transport the sniper team to its insertion site. Wheeled or tracked vehicle insertion requires the same planning considerations used in other insertion techniques. The team risks compromise if it uses vehicle insertion beyond the FLOT due to noise. Enemy OPs and scout elements can easily detect and prevent infiltration of the sniper team. However, this technique can be effectively used in support of immediate battle operations by using deceptive measures.

Section II
EXECUTION

The execution phase consists of movement from the insertion site to the target area, mission execution, and movement to the extraction site.

6-6. MOVEMENT TO TARGET AREA

After leaving the insertion site, the sniper team transmits an initial entry report as required by unit SOP. This report ensures operable radio equipment and provides the team's status at the same time.

a. **Route Selection.** No matter which means of insertion, the selection of the route to the target area is critical.

(1) Enemy location, detection devices, and defensive capabilities; terrain; weather; and man-made obstacles are all to be considered when selecting the primary and alternate routes. En route checkpoints are selected to keep track of the team.

(2) The team uses NODs to operate during reduced visibility. The team's extensive training and land navigation skills allow it to rapidly traverse rugged terrain and to avoid detection.

b. **Movement Interval.** The interval between sniper team members may vary during movement into the target area. It is based on visibility, terrain, and enemy disposition. The team keys movement to the following rules, which should be discussed in detail in the sniper SOP.

(1) Maintain visual contact at a normal interval. (Intervals can expand and contract based on terrain and visibility.)

(2) Always maintain noise and light discipline.

(3) Observe the assigned sector of responsibility.

(4) React together (for example, when one gets down, they both get down.)

(5) Ensure the sniper team leader positions himself to the rear of the observer.

(6) Move on routes that best conceal movement from enemy observation and cover movement from direct enemy fire.

(7) Ensure the interval between members closes when moving through obstructions (darkness, smoke, heavy brush, narrow passes, and mine fields); ensure the interval opens when obstructions to movement and control lessen.

c. **Movement Security.** Each sniper team member must be security conscious, maintaining constant all-round security. During movement, each team member is responsible for an assigned security sector. The sniper team's route makes the best use of cover and concealment, and security or listening halts are made, as needed. Personal and equipment camouflage is enforced at all times.

d. **Arm-and-Hand Signals.** The sniper establishes standard arm-and-hand signals to reduce oral communications and to assist in control. These signals should conform to those listed in FM 21-75 and the sniper SOP.

6-7. OCCUPATION OF POSITION

The tentative final firing position, ORP, and route are selected during the mission planning phase by map and aerial photograph reconnaissance. The sniper team moves close to the TFFP and sets up an objective rally point. It then moves forward to search for a TFFP, ensuring the site is suitable and the target area can be observed at ground level. At this point, the TFFP becomes an FFP. Reconnaissance should be made during limited visibility. The team returns to the ORP, secures all mission-essential equipment, and moves to the FFP and occupies it. The sniper team watches and listens for the enemy before constructing the hide position (METT-T dependent).

6-8. SITE SELECTION

Selection of the firing position is METT-T dependent. As a minimum, the sniper team uses the following criteria when selecting an FFP:

a. Ensures that an unrestricted observation of the target area is possible. The team can then place the designated target area under constant, effective surveillance and within the range of RSTA devices and the sniper's weapon system.

b. Selects an area that provides a concealed entrance and exit routes.

c. Avoids man-made objects.

d. Avoids dominant or unusual terrain features.

e. Selects an area that is dry, or has good drainage and is not prone to flooding.

f. Selects an area that the enemy would not occupy.

g. Avoids the skyline or blending backgrounds.

h. Avoids roads or trails.

i. Avoids natural lines of movement (gullies, draws, or any terrain that affords easy foot movement).

j. Selects an area in which the team cannot be easily trapped.

k. Ensures it has a natural obstacle to vehicles between the FFP and the target area, if possible (roadside ditch, fence, wall, stream, or river).

1. Selects an area downwind of inhabited areas, if possible.

m. Selects an area in or near a suitable communications site.

n. Avoids the normal line of vision of the enemy in the target area.

o. Selects an area near a source of water.

6-9. REPORTS

The sniper team follows the communications procedures as outlined in the unit SOP. The team members must ensure that communications are maintained throughout the mission by the use of directional antennas, masking, and burst transmissions.

a. The sniper team does not analyze information it only collects and reports based on SIR. The team must format information reporting IAW the unit SOP and the type of communications equipment used.

b. Other reports that the sniper team may use, such as emergency resupply, communication checks, and emergency extraction, should also be formatted IAW the SOP.

6-10. MOVEMENT TO EXTRACTION SITE

Movement to a planned extraction site will be necessary in many operations. The sniper team must observe the principles of route selection and movement security.

a. **Priorities.** The time that a sniper team remains beyond the FFL depends on its mission and equipment. The extraction is critical from a standpoint of morale and mission accomplishment. Plans for extraction by air, ground, or water are made before the operation, with alternate plans for contingencies such as the evacuation of sick or injured personnel. During the mission, the sniper may be faced with an unforeseen situation that may demand the utmost in flexibility, discipline, and leadership.

b. **Code Words.** Each sniper team is given code words in the OPORD for use during extraction. For example, one code word may mean that the team is at its pickup zone. Another may mean that both the primary and alternate pickup zones are compromised and to abort the extraction.

c. **No Communication.** When a sniper team has missed a certain number of required transmissions, the operations section assumes that the team has a communications problem, is in trouble, or both. At that time, the no-communication extraction plan is used.

d. **Alternatives.** Extraction of the sniper team may be by means other than air. The OPORD may specify to extract the team by land or water, or to link up with friendly forces in an offensive operation. Any of these means may also be planned as alternates to avoid capture or if the sniper team cannot be extracted by air.

e. **Ground Extraction.** Despite the desirability of extracting the team by aircraft or linkup, use of these methods may be prevented by security of the sniper team, poor communications, or enemy air defense. The sniper team must be thoroughly trained in exfiltration techniques so they can walk out, either one at a time or together.

Section III
EXTRACTION AND RECOVERY

The sniper team performs an extraction as quickly as possible after the mission is accomplished. An extraction site is always planned and coordinated with supporting forces. However, the situation may dictate that the sniper decides whether to use the planned extraction site or to exfiltrate.

6-11. PLANNING

The sniper team must be prepared to exfiltrate over predetermined land routes to friendly lines as a team (or individually) or to exfiltrate to an area for extraction by air or water. Planning includes the following:

a. **Distance.** Distance may prevent an all-land exfiltration. The initial phase may be by land, ending in extraction by air or water.

b. **Terrain.** The terrain is important in selecting extraction means. The extraction site must offer favorable tactical considerations, tide data, PZ suitability, and cover from enemy direct-fire weapons. The sniper team uses the most unlikely terrain for extraction such as swamps, jungles, and mountain areas.

c. **Enemy.** Enemy pressure can develop during the extraction. Detailed plans must be made for contingency exfiltrations forced by the enemy.

d. **Evasion and Escape.** Preinsertion planning must include the development of a viable evasion and escape plan. The sniper team must do the following

(1) Checks all factors that deal with survival and evasion opportunities.

(2) Devises an evasion and escape plan that provides the best chance of survival and return to friendly lines in view of the hazards involved and mission objectives.

(3) Becomes familiar with the evasion and escape plans.

6-12. EVASION AND ESCAPE PLAN

Each mission has its specific problems associated with evasion and escape. The plan must conform to these unique problems while exploiting individual abilities, training of sniper team members, and supporting air or boat crews. The following general rules apply to evasion and escape plans for sniper operations:

a. The purpose of the plan is to attempt to save the individual who can no longer complete the assigned mission.

b. When sniper teams are behind enemy lines, the most successful escapes may involve air or water movement away from enemy-held territory.

c. Evasion and escape plans involve the following three phases:

(1) Phase one occurs during entry into the target area.

(2) Phase two occurs near the target area. It allows the sniper team to pursue its mission with a reasonable chance of success.

(3) Phase three occurs after the mission is accomplished. It is often the most difficult time to evade and escape.

d. The sniper team may be required to hide for several days to allow the enemy to become complacent before the team tries to move.

e. In selecting extraction sites, the sniper considers the danger of compromising other activities. He must prepare alternate plans for unforeseen developments.

6-13. AIR OR WATER EXTRACTION

Extraction by air or water is favored when resources are available and when it will not compromise the mission.

a. Other considerations that favor this method areas follows:

(1) Long distances must be covered.

(2) The time of return is essential.

(3) The enemy does not have air and naval superiority.

(4) Heavily populated hostile areas obstruct exfiltration.

(5) The team cannot be resupplied.

(6) Casualties must be extracted.

b. Several techniques maybe used to extract the team.

(1) Helicopter landing is the best method since the sniper team and its equipment can board the helicopter quickly.

(2) The troop ladder is the second best method. It lets sniper team members board the helicopter, but the helicopter can liftoff while snipers are still on the ladder.

(3) The STABO extraction system allows rapid pickup of one to four soldiers, who are suspended on lines beneath the helicopter. Soldiers are picked up and moved to an area where the helicopter can land. The sniper team then boards the helicopter.

(4) The jungle penetrator retrieves soldiers from areas where helicopters cannot land. It can pickup 1 to 3 persons at a time.

(5) The SPIES can extract soldiers from areas where helicopters cannot land. It can pickup 1 to 10 soldiers at a time.

6-14. LAND EXFILTRATION

This method is favored when snipers are not too far from friendly lines or no other means of extraction is available. It is also used when the terrain provides cover and concealment for foot movement and limits the employment of enemy mobile units against the exfiltrating team. Other considerations favoring this method are as follows:

a. Areas along exfiltration routes are uninhabited.

b. The enemy force is widely dispersed or is under such pressure that it is difficult for them to concentrate against the exfiltrating team.

c. The enemy force can stop an air or water extraction.

6-15. VEHICLE EXTRACTION

Vehicle extraction involves the exfiltration of the sniper team to an extraction site for extraction by a wheeled or tracked vehicle. Planning and coordination must be made during the preinsertion phase. Contingency plans must also be made to avoid compromise or any unforeseen situations.

6-16. RECOVERY

Recovery is the last phase of a sniper operation. It consists of the sniper team's return to the operations base, debriefing, equipment maintenance and turn-in, and stand-down. At the end of this phase, the sniper team prepares for future missions. (See Chapter 5.)

CHAPTER 7

COMMUNICATIONS

The basic requirement of combat communications is to provide rapid, reliable, and secure interchange of information.

Section I
FIELD-EXPEDIENT ANTENNAS

Communications are a vital aspect in successful mission accomplishment. The information in this section helps the sniper team maintain effective communications and correct any radio antenna problems.

7-1. REPAIR TECHNIQUES

Antennas are sometimes broken or damaged, causing either a communications failure or poor communications. If a spare antenna is available, the damaged antenna is replaced. When there is no spare, the sniper team may have to construct an emergency antenna. The following paragraphs contain suggestions for repairing antennas and antenna supports and the construction and adjustment of emergency antennas.

DANGER
SERIOUS INJURY OR DEATH CAN RESULT FROM CONTACT WITH THE RADIATING ANTENNA OF A MEDIUM-POWER OR HIGH-POWER TRANSMITTER. TURN THE TRANSMITTER OFF WHILE MAKING ADJUSTMENTS TO THE ANTENNA.

---

a. **Whip Antennas.** When a whip antenna is broken into two sections, the part of the antenna that is broken off can be connected to the part attached to the base by joining the sections. (Use the method shown in A, Figure 7-1, when both parts of the broken whip are available and usable.) (Use the method in B, Figure 7-1, when the part of the whip that was broken off is lost or when the whip is so badly damaged that it cannot be used.) To restore the antenna to its original length, a piece of wire is added that is nearly the same length as the missing part of the whip. The pole support is then lashed securely to both sections of the antenna. The two antenna sections are cleaned thoroughly to ensure good contact before connecting them to the pole support. If possible, the connections are soldered.

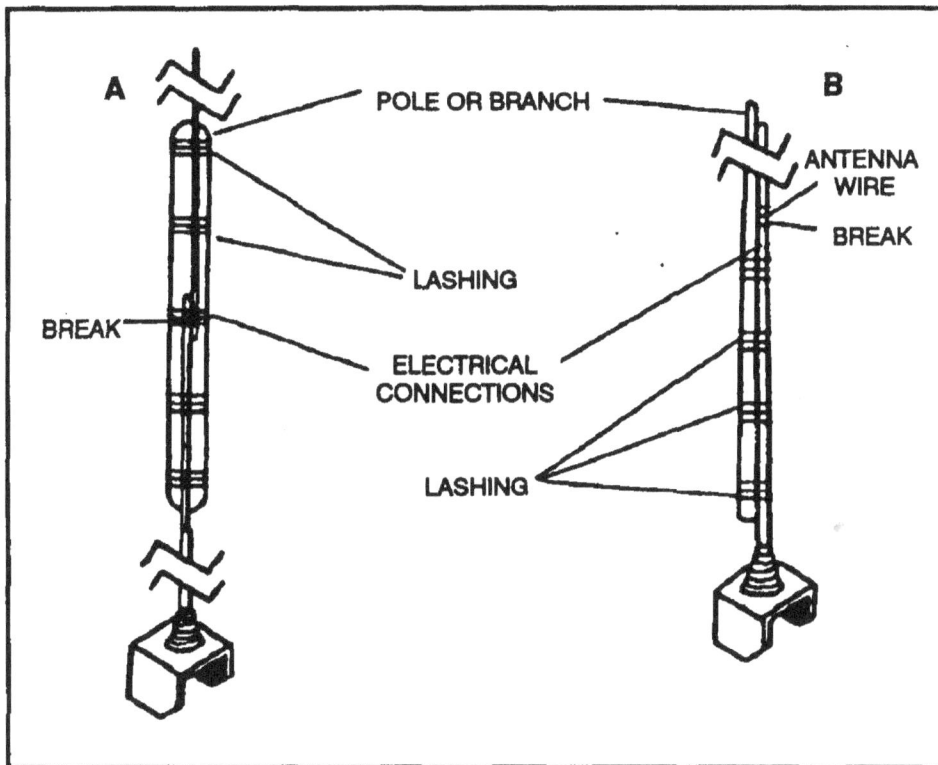

Figure 7-1. Emergency repair of broken whip antenna.

b. **Wire Antennas.** Emergency repair of a wire antenna may involve the repair or replacement of the wire used as the antenna or transmission line; or the repair or replacement of the assembly used to support the antenna.

(1) When one or more wires of an antenna are broken, the antenna can be repaired by reconnecting the broken wires. To do this, lower the antenna to the ground, clean the ends of the wires, and twist the wires together. Whenever possible, solder the connection.

(2) If the antenna is damaged beyond repair, construct a new one. Make sure that the length of the wires of the substitute antenna are the same length as those of the original.

(3) Antenna supports may also require repair or replacement. A substitute item may be used in place of a damaged support and, if properly insulated, can be of any material of adequate strength. If the radiating element is not properly insulated, field antennas may be shorted to ground and be ineffective. Many commonly found items can be used as field-expedient insulators. The best of these items are plastic or glass to include plastic spoons, buttons, bottle necks, and plastic bags. Though less effective than plastic or glass but still better than no insulator at all are wood and rope. The radiating element—the actual antenna wire-should touch only the antenna terminal and should be physically separated from all other objects, other than the supporting insulator. (See Figure 7-2 for various methods of making emergency insulators.)

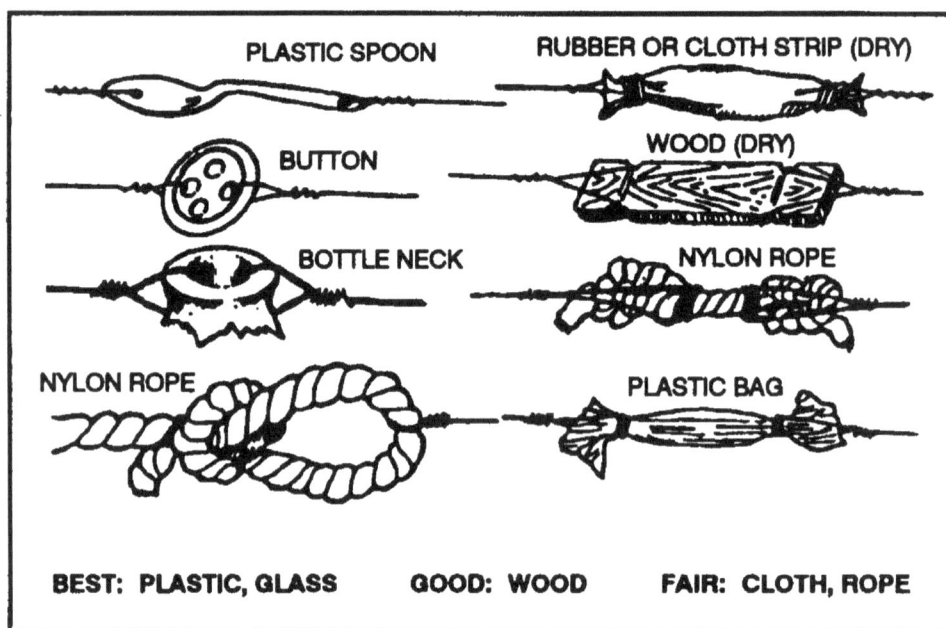

Figure 7-2. Improvised Insulators.

7-2. CONSTRUCTION AND ADJUSTMENT
Sniper teams may use the following methods to construct and adjust antennas.

a. **Construction.** The best kinds of wire for antennas are copper and aluminum. In an emergency, however, snipers use any type of wire that is available.

(1) The exact length of most antennas is critical. The emergency antenna should be the same length as the antenna it replaces.

(2) Antennas supported by trees can usually survive heavy wind storms if the trunk of a tree or a strong branch is used as a support. To keep the antenna taut and to prevent it from breaking or stretching as the trees sway, the sniper attaches a spring or old inner tube to one end of the antenna. Another technique is to pass a rope through a pulley or eyehook. The rope is attached to the end of the antenna and loaded with a heavyweight to keep the antenna tightly drawn.

(3) Guidelines used to hold antenna supports are made of rope or wire. To ensure the guidelines will not affect the operation of the antenna, the sniper cuts the wire into several short lengths and connects the pieces with insulators.

b. **Adjustment.** An improvised antenna may change the performance of a radio set. The following methods can be used to determine if the antenna is operating properly

(1) A distant station may be used to test the antenna. If the signal received from this station is strong, the antenna is operating satisfactorily. If the signal is weak, the sniper adjusts the height and length of the antenna and the transmission line to receive the strongest signal at a given setting on the volume control of the receiver. This is the best method of tuning an antenna when transmission is dangerous or forbidden.

(2) In some radio sets, the sniper uses the transmitter to adjust the antenna. First, he sets the controls of the transmitter to normal; then, he tunes the system by adjusting the antenna height, the antenna length, and the transmission line length to obtain the best transmission output.

7-3. FIELD-EXPEDIENT OMNIDIRECTIONAL ANTENNAS

Vertical antennas are omnidirectional. The omnidirectional antenna transmits and receives equally well in all directions. Most tactical antennas are vertical; for example, the man-pack portable radio uses a vertical whip and so do the vehicular radios in tactical vehicles. A vertical antenna can be made by using a metal pipe or rod of the correct length, held erect by means of guidelines. The lower end of the antenna should be insulated from the ground by placing it on a large block of wood or other insulating material. A vertical antenna may also be a wire supported by a tree or a wooden pole (Figure 7-3). For short vertical antennas, the pole may be used without guidelines (if properly supported at the base). If the length of the vertical mast is not long enough to support the wire upright, it may be necessary to modify the connection at the top of the antenna (Figure 7-4). (See FM 24-18.)

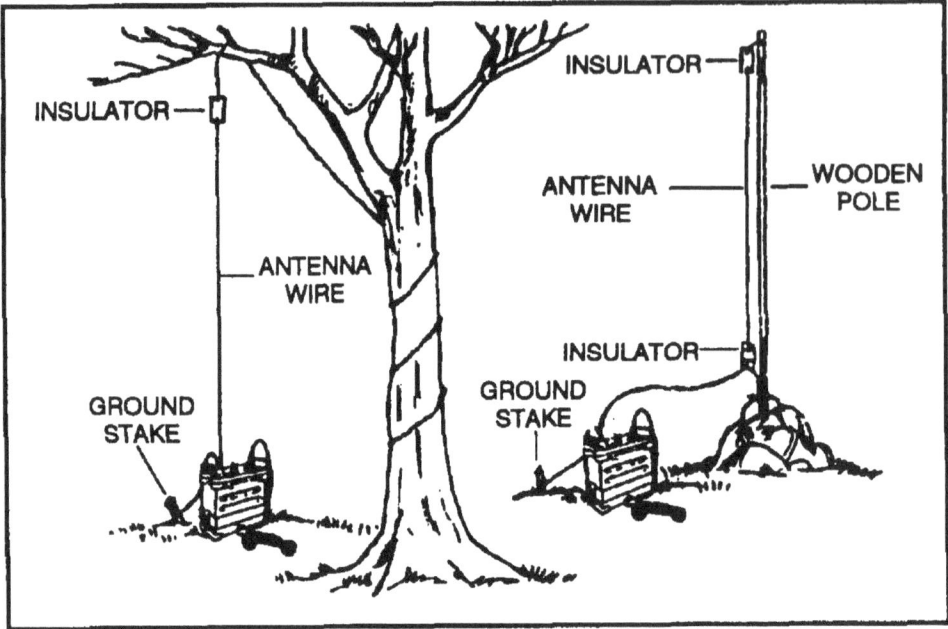

Figure 7-3. Field substitutes for support of vertical wire antennas.

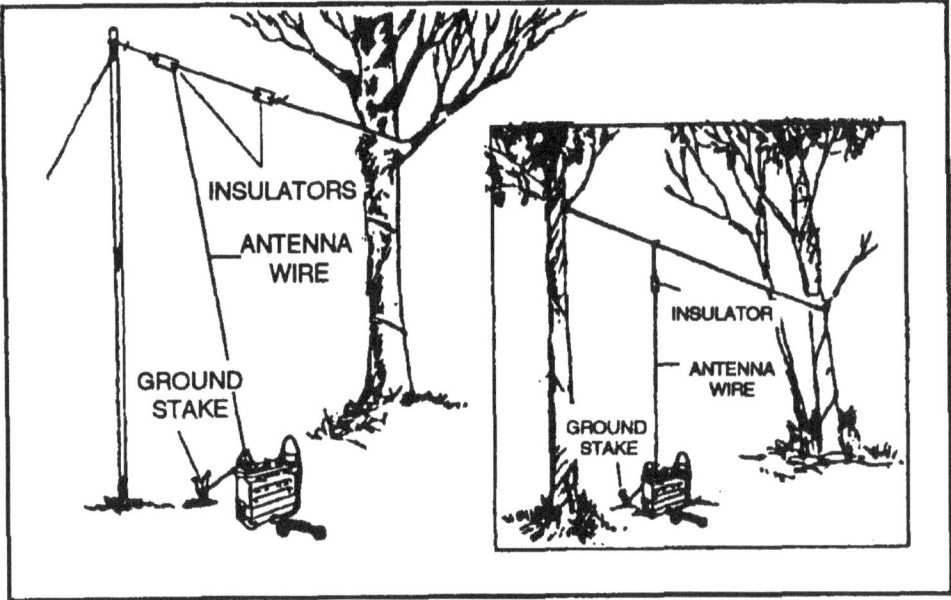

Figure 7-4. Additional means of supporting vertical wire antennas.

a. **End-Fed Half-Wave Antenna.** An emergency, end-fed half-wave antenna (Figure 7-5, page 7-6) can be constructed from available materials such as field wire, rope, and wooden insulators. The electrical length of this antenna is measured from the antenna terminal on the radio set to the far end of the antenna. The best performance can be obtained by constructing the antenna longer than necessary and then shortening it, as required, until the best results are obtained. The ground terminal of the radio set should be connected to a good earth ground for this antenna to function efficiently.

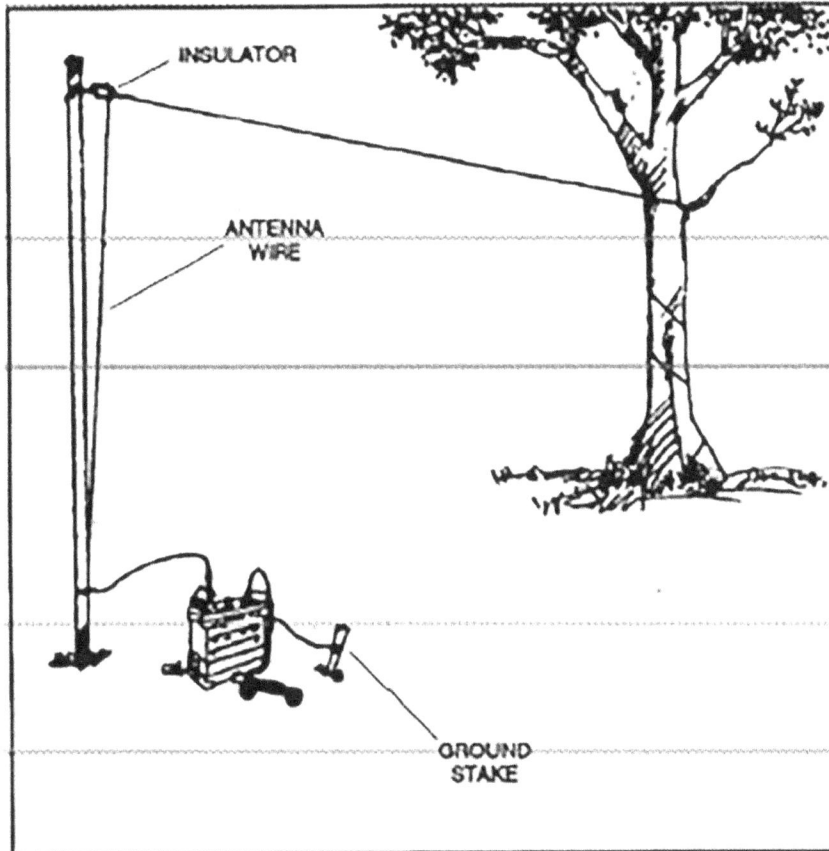

Figure 7-5. End-fed half-wave antenna.

b. Center-Fed Doublet Antenna. The center-fed doublet is a half-wave antenna consisting of two quarter wavelength sections on each side of the center (Figure 7-6). Doublet antennas are directional broadside to their length, which makes the vertical doublet antenna omnidirectional. This is because the radiation pattern is doughnut-shaped and bidirectional.

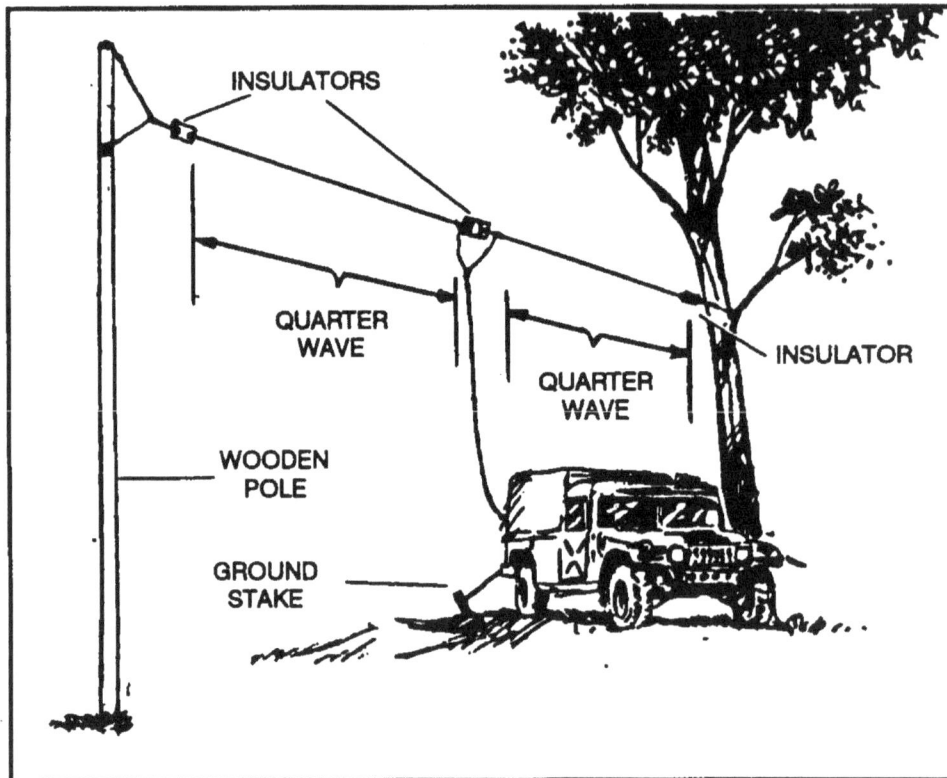

Figure 7-6. Center-fed half-wave doublet antenna.

(1) Compute the length of a half-wave antenna by using the formula in paragraph 7-5. Cut the wires as close as possible to the correct length; this is very important.

(2) Uses transmission line for conducting electrical energy from one point to another and for transferring the output of a transmitter to an antenna. Although it is possible to connect an antenna directly to a transmitter, the antenna is usually located some distance away.

(3) Support center-fed half-wave FM antennas entirely with pieces of wood. (A horizontal antenna of this type is shown in A, Figure 7-7, page 7-8, and a vertical antenna in B, Figure 7-7.) Rotate these antennas to any position to obtain the best performance.

(a) If the antenna is erected vertically, bring out the transmission line horizontally from the antenna for a distance equal to at least one-half of the antenna's length before it is dropped down to the radio set.

(b) The half-wave antenna is used with FM radios (Figure 7-8, page 7-8). It is effective in heavily wooded areas to increase the range of portable radios. Connect the top guidelines to a limb or pass it over the limb and connect it to the tree trunk or a stake.

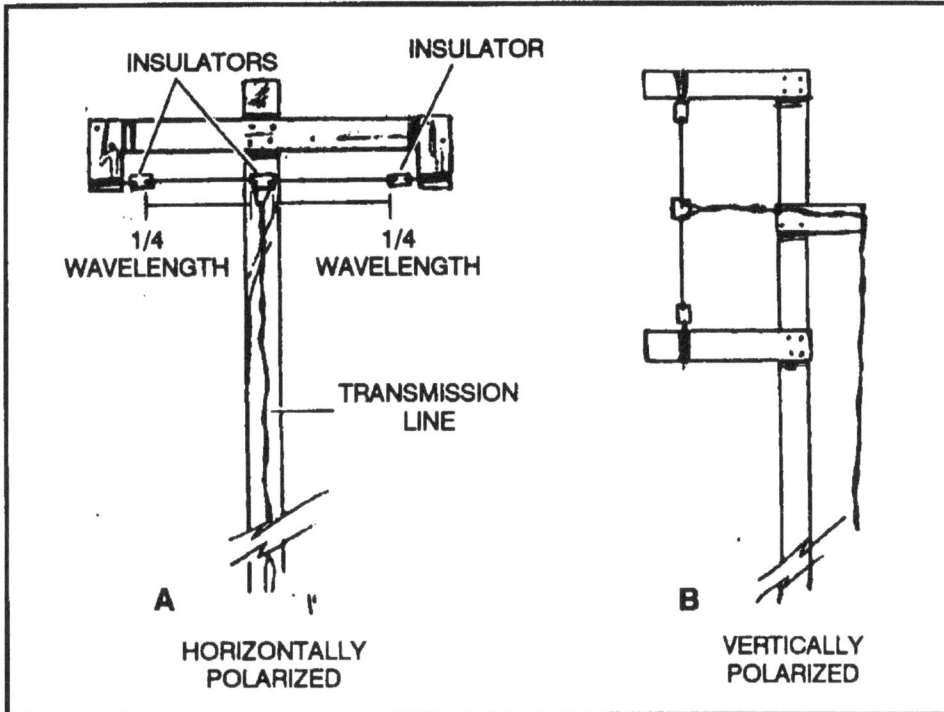

Figure 7-7. Center-fed half-wave antenna, supported.

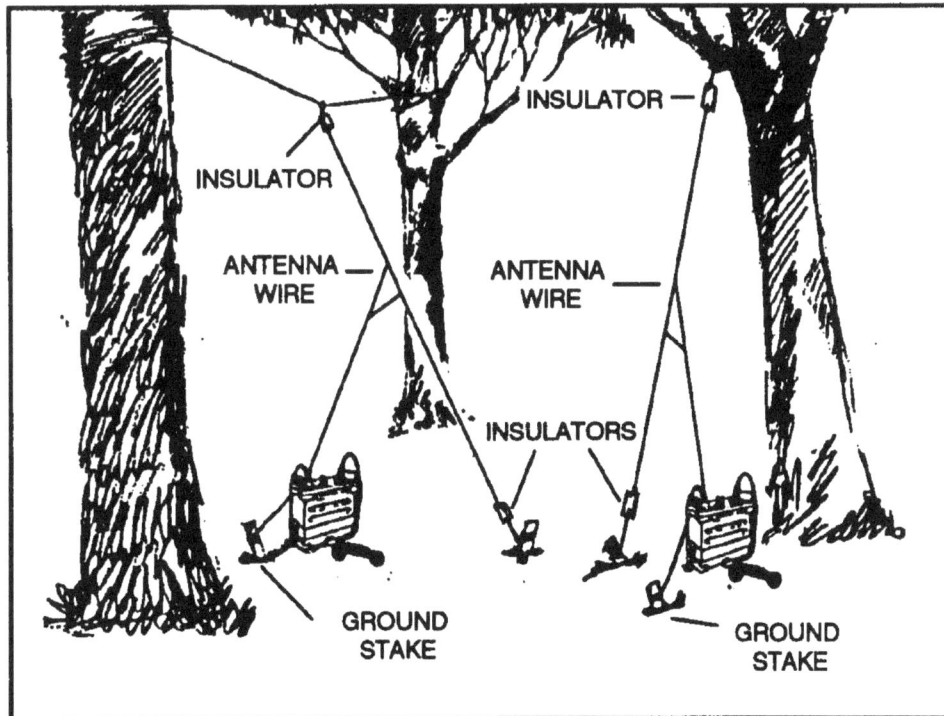

Figure 7-8. Improvised vertical half-wave antenna.

7-4. FIELD-EXPEDIENT DIRECTIONAL ANTENNAS

The vertical half-rhombic antenna (Figure 7-9) and the long-wire antenna (Figure 7-10) are two field-expedient directional antennas. These antennas consist of a single wire, preferably two or more wavelengths long, supported on poles at a height of 3 to 7 meters (10 to 20 feet) above the ground. The antennas will, however, operate satisfactorily as low as 1 meter (about 3 feet) above the ground—the radiation pattern is directional. The antennas are used mainly for either transmitting or receiving high-frequency signals.

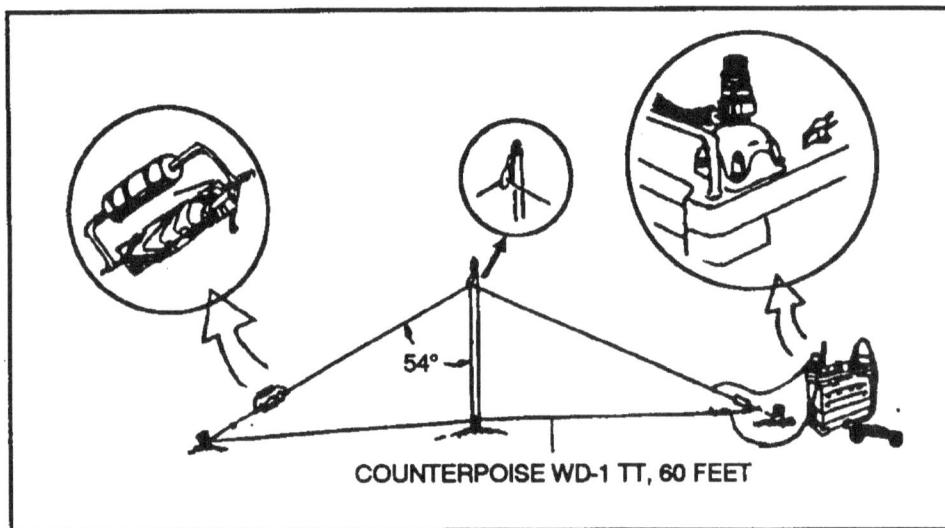

COUNTERPOISE WD-1 TT, 60 FEET

Figure 7-9. Verticle half-rhombic antenna.

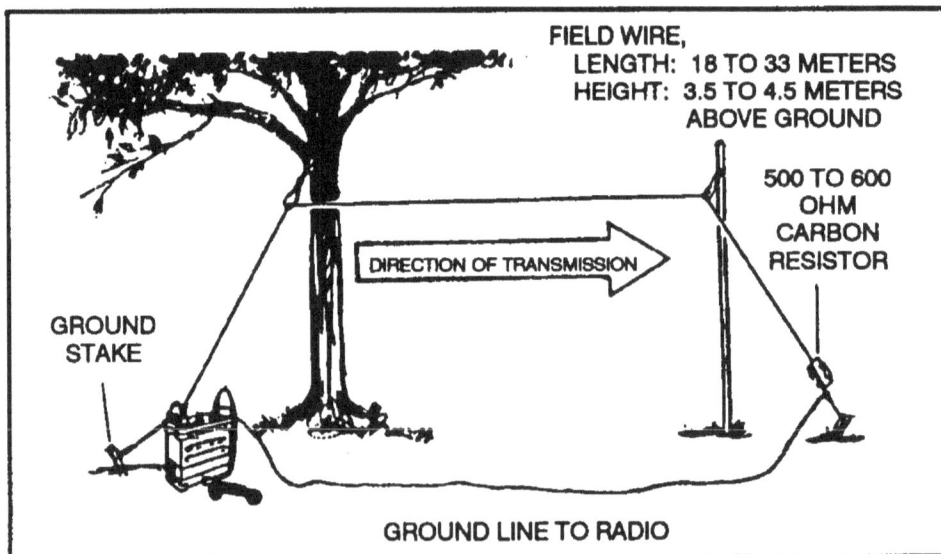

FIELD WIRE,
LENGTH: 18 TO 33 METERS
HEIGHT: 3.5 TO 4.5 METERS
ABOVE GROUND

500 TO 600 OHM CARBON RESISTOR

DIRECTION OF TRANSMISSION

GROUND STAKE

GROUND LINE TO RADIO

Figure 7-10. Long-wire antenna.

a. The V antenna (Figure 7-11) is another field-expedient directional antenna. It consists of two wires forming a V with the open area of the V pointing toward the desired direction of transmission or reception. To make construction easier, the legs should slope downward from the apex of the V; this is called a *sloping-V antenna* (Figure 7-12). The angle between the legs varies with the length of the legs to achieve maximum performance. (to determine the angle and the length of the legs, use the table in Table 7-l.)

b. When the antenna is used with more than one frequency or wavelength, use an apex angle that is midway between the extreme angles determined by the chart. To make the antenna radiate in only one direction, add noninductive terminating resistors from the end of each leg (not at the apex) to ground. (See TM 11-666.)

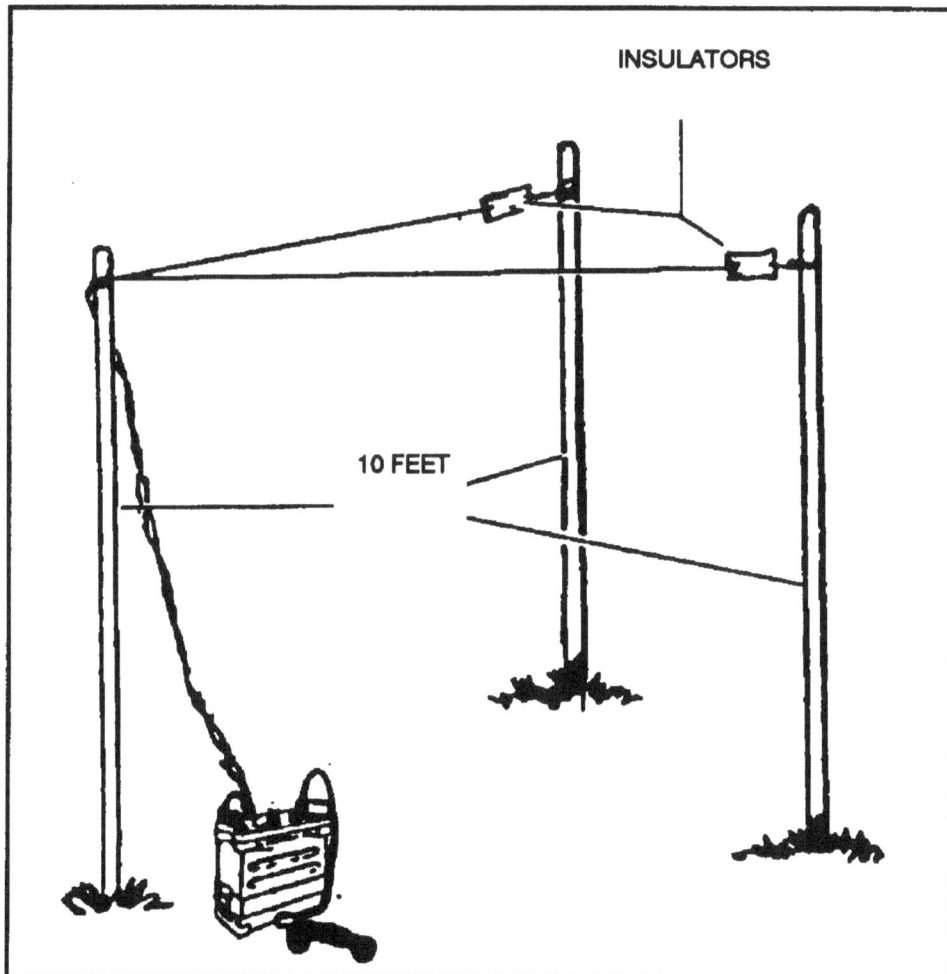

Figure 7-11. V antenna.

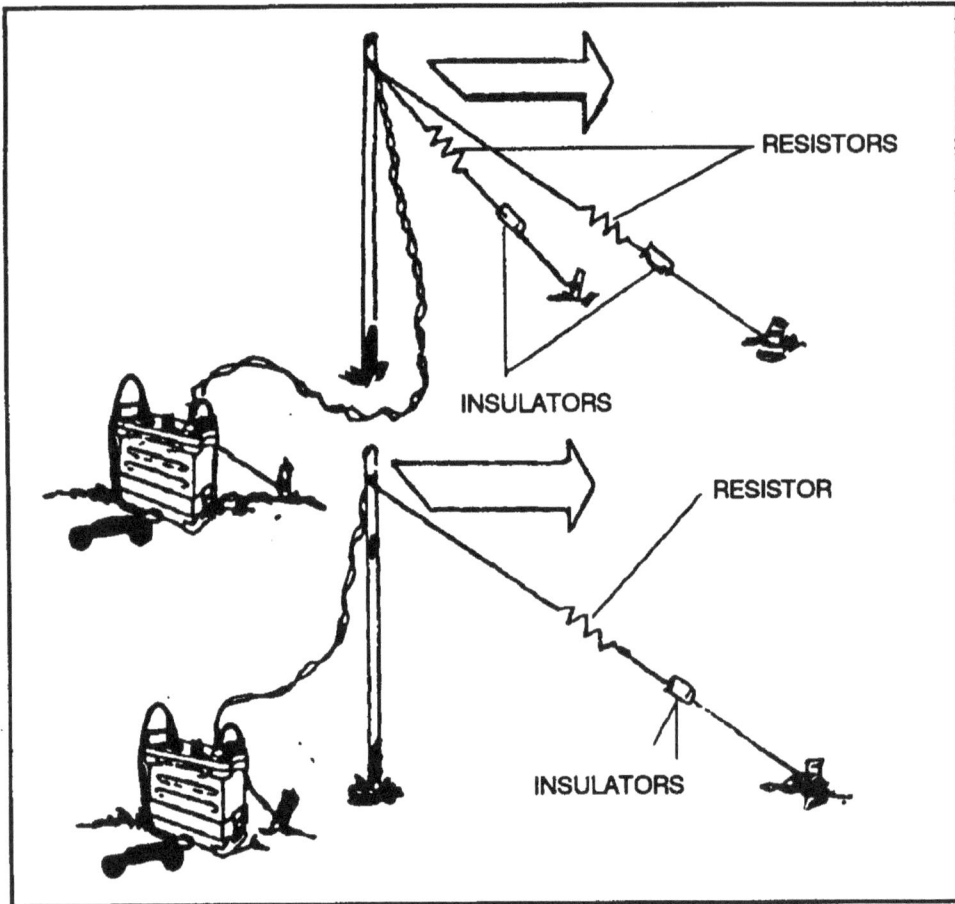

Figure 7-12. Sloping-V antenna.

ANTENNA LENGTH (wavelength)	OPTIMUM APEX ANGLE (degrees)
1	90
2	70
3	58
4	50
6	40
8	35
10	33

Table 7-1. Leg angle for V antennas.

7-5. ANTENNA LENGTH

The length of an antenna must be considered in two ways: both a physical and an electrical length. These two lengths are never the same. The reduced velocity of the wave on the antenna and a capacitive effect (known as end effect) make the antenna seem longer electrically than it is physically. The contributing factors are the ratio of the diameter of the antenna to its length and the capacitive effect of terminal equipment, such as insulators and clamps, used to support the antenna.

a. To calculate the physical length of an antenna, use a correction of 0.95 for frequencies between 3.0 and 50.0 MHz The figures given below are for a half-wave antenna.

$$\text{Length (meters)} = \frac{150 \ \times \ 0.95}{\text{Frequency in MHz}} = \frac{142.5}{\text{Frequency in MHz}}$$

$$\text{Length (feet)} = \frac{492 \ \times \ 0.95}{\text{Frequency in MHz}} = \frac{468}{\text{Frequency in MHz}}$$

b. Use the following formula to calculate the length of a long-wire antenna (one wavelength or longer) for harmonic operation:

$$\text{Length (meters)} = \frac{150 \ (N-0.05)}{\text{Frequency in MHz}}$$

$$\text{Length (feet)} = \frac{492 \ (N-0.05)}{\text{Frequency in MHz}}$$

N equals the number of half-wavelengths in the total length of the antenna. For example, if the number of half-wavelengths is 3 and the frequency in MHz is 7, then—

$$\text{Length (meters)} = \frac{150(N-0.05)}{\text{Frequency in MHz}} = 1 \frac{50(3-.05)}{7} =$$

$$\frac{150 \times 2.95}{7} = \frac{442.50}{7} = 63.2 \text{ meters}$$

7-6. ANTENNA ORIENTATION

If the azimuth of the radio path is not provided, the azimuth should be determined by the best available means. The accuracy required in determining the azimuth of the path depends on the radiation pattern of the directional antenna. In transportable operation, the rhombic and V antennas may have such a narrow beam as to require great accuracy in azimuth determination. The antenna should be erected for the correct azimuth. Great accuracy is not required in erecting broad-beam antennas. Unless a line of known azimuth is available at the site, the direction of the path is best determined by a magnetic compass.

7-7. IMPROVEMENT OF MARGINAL COMMUNICATIONS

Under certain situations, it may not be feasible to orient directional antennas to the correct azimuth of the desired radio path. As a result, marginal communications may suffer. To improve marginal communications, the following procedure can be used:

    a. Check, tighten, and tape cable couplings and connections.

    b. Return all transmitters and receivers in the circuit.

    c. Ensure antennas are adjusted for the proper operating frequency.

    d. Change the heights of antennas.

    e. Move the antenna a short distance away and in different locations from its original location.

Section II
RADIO OPERATIONS UNDER UNUSUAL CONDITIONS

The possibility of being deployed to different parts of the world presents many problems for the sniper team due to extremes in climate and terrain. This section informs the sniper team of these common problems and possible solutions to eliminate or reduce adverse effects.

7-8. ARCTIC AREAS

Single-channel radio equipment has certain capabilities and limitations that must be carefully considered when operating in cold areas. However, in spite of limitations, radio is the normal means of communications in such areas. One of the most important capabilities of the radio in Arctic-like areas is its versatility. Man-packed radios can be carried to any point accessible by foot or aircraft. A limitation on radio communications that radio operators must expect in extremely cold areas is interference by ionospheric disturbances. These disturbances, known as ionospheric storms, have a definite degrading effect on skywave propagation. Moreover, either the storms or the auroral (such as northern lights) activity can cause complete failure of radio communications. Some frequencies may be blocked completely by static for extended periods during storm activity. Fading, caused by changes in the density and height of the ionosphere, can also occur and may last from minutes to weeks. The occurrence of these disturbances is difficult to predict. When they occur, the use of alternate frequencies and a greater reliance on FM or other means of communications are required.

a. **Antenna Installation.** Antenna installation in Arctic-like areas presents no serious problems. However, installing some antennas may take longer because of adverse working conditions. Some suggestions for installing antennas in extremely cold areas areas follows:

(1) Antenna cables must be handled carefully since they become brittle in low temperatures.

(2) Whenever possible, antenna cables should be constructed overhead to prevent damage from heavy snow and frost. Nylon rope guidelines, if available, should be used in preference to cotton or hemp because nylon ropes do not readily absorb moisture and are less likely to freeze and break.

(3) An antenna should have extra guidelines, supports, and anchor stakes to strengthen it to withstand heavy ice and wind.

(4) Some radios (usually older generation radios) adjusted to a specific frequency in a relatively warm place may drift off frequency when exposed to extreme cold. Low battery voltage can also cause frequency drift. When possible, a radio should warmup several minutes before placing it into operation. Since extreme cold tends to lower output voltage of a dry battery, warming the battery with body heat before operating the radio set can reduce frequency drift.

(5) Flakes or pellets of highly electrically charged snow is sometimes experienced in northern regions. When these particles strike the antenna, the resulting electrical discharge causes a high-pitched static roar that can blanket all frequencies. To overcome this static, antenna elements can be covered with polystyrene tape and shellac.

b. **Maintenance Improvement in Arctic Areas.** The maintenance of radio equipment in extreme cold presents many problems. Radio sets must be protected from blowing snow since snow will freeze to dials and knobs and blow into the wiring to cause shorts and grounds. Cords must be handled carefully as they may lose their flexibility in extreme cold. All radio equipment must be properly winterized. The appropriate technical manual should be checked for winterization procedures. Some suggestions for maintenance in Arctic areas include:

(1) *Batteries.* The effect of cold weather conditions on wet and dry cell batteries depends on the following factors: the type and kind of

battery, the load on the battery, the specific use of the battery, and the degree of exposure to cold temperatures.

(2) *Winterization.* The radio set technical manual should rechecked for special precautions for operation in extremely cold climates. For example, normal lubricants may solidify and cause damage or malfunctions. They must be replaced with the recommended Arctic lubricants.

(3) *Microphone.* Moisture from the sniper's breath may freeze on the perforated cover plate of his microphone. Standard microphone covers can be used to prevent this. If standard covers are not available, a suitable cover can be improvised from rubber or cellophane membranes or from rayon or nylon cloth.

(4) *Breathing and sweating.* A radio set generates heat when it is operated. When turned off, the air inside the radio set cools and contracts, and draws cold air into the set from the outside. This is called *breathing*. When a radio breathes and the still-hot parts come in contact with subzero air, the glass, plastic, and ceramic parts of the set may cool too rapidly and break. When cold equipment is brought suddenly into contact with warm air, moisture condenses on the equipment parts. This is called *sweating*. Before cold equipment is brought into a heated area, it should be wrapped in a blanket or parka to ensure that it warms gradually to reduce sweating. Equipment must be thoroughly dry before it is taken into the cold air or the moisture will freeze.

7-9. JUNGLE AREAS

Radio communications in jungle areas must be carefully planned, because the dense jungle growth reduces the range of radio transmission. However, since single-channel radio can be deployed in many configurations, especially man-packed, it is a valuable communications asset. The capabilities and limitations of single-channel radio must be carefully considered when used by forces in a jungle environment. The mobility and various configurations in which a single-channel radio can be deployed are its main advantages in jungle areas. Limitations on radio communications in jungle areas are due to the climate and the density of jungle growth. The hot and humid climate increases maintenance problems of keeping the equipment operable. Thick jungle growth acts as a vertically polarized absorbing screen for radio frequency energy that, in effect, reduces transmission range. Therefore, increased emphasis on maintenance and antenna siting is a must when operating in jungle areas.

a. **Jungle Operational Techniques.** The main problem in establishing radio communications in jungle areas is the siting of the antenna.

The following techniques can be applied to improve communications in the jungle:

(1) Locate antennas in clearings on the edge farthest from the distant station and as high as possible.

(2) Keep antenna cables and connectors off the ground to lessen the effects of moisture, fungus, and insects. This also applies to all power and telephone cables.

(3) Use complete antenna systems, such as ground planes and dipoles, for more effect than fractional wavelength whip antennas.

(4) Clear vegetation from antenna sites. If an antenna touches any foliage, especially wet foliage, the signal will be grounded.

(5) When wet, vegetation acts like a vertically polarized screen and absorbs much of a vertically polarized signal. Use horizontally polarized antennas in preference to vertically polarized antennas.

b. **Maintenance Improvement in the Jungle.** Due to moisture and fungus, the maintenance of radio sets in tropical climates is more difficult than intemperate climates The high relative humidity causes condensation to form on the equipment and encourages the growth of fungus. Operators and maintenance personnel should check appropriate technical manuals for special maintenance requirements. Some techniques for improving maintenance in jungle areas follow:

(1) Keep the equipment as dry as possible and in lighted areas to retard fungus growth.

(2) Clear all air vents of obstructions so air can circulate to cool and dry the equipment.

(3) Keep connectors, cables, and bare metal parts as free of fungus growth as possible.

(4) Use moisture and fungus-proofing paint to protect equipment after repairs are made or when equipment is damaged or scratched.

c. **Expedient Antennas.** Sniper teams can improve their ability to communicate in the jungle by using expedient antennas. While moving, the team is usually restricted to using the short and long antennas that come with the radios. However, when not moving, snipers can use these expedient antennas to broadcast farther and to receive more clearly. However, an antenna that is not "tuned" or "cut" to the operating frequency is not as effective as the whips that are supplied with the radio. Circuits inside the radio "load" the whips properly so that they are "tuned" to give the greatest output. Whips are not as effective as a tuned doublet or tuned ground plane (namely RC 292-type), but the doublet or ground plane must be tuned to the operating frequency. This is especially critical with low-power radios such as the AN/ PRC-77.

(1) *Expedient 292-type antenna.* The expedient 292-type antenna was developed for use in the jungle and, if used properly, can increase the team's ability to communicate. In its entirety, the antenna is bulky, heavy, and not acceptable for sniper team operations. The team can, however, carry only the mast head and antenna sections, mounting these on wood poles or hanging them from trees; or, the team can make a complete expedient 292-type antenna (Figure 7-13, page 7-18), using WD-1, wire, and other readily available material. The team can also use almost any plastic, glass, or rubber objects for insulators. Dry wood is acceptable when nothing else is available. (See Figure 7-2 for types of insulators that may be used.) The following describes how to make this antenna:

(a) Use the quick-reference table (Table 7-2, page 7-19) to determine the length of the elements (one radiating and three ground planes) for the frequency that will be used. Cut these elements (A, Figure 7-13, page 7-18) from WD-1 field wire (or similar wire). Cut spacing sticks (B, Figure 7-13) the same length. Place the ends of the sticks together to form a triangle and tie the ends with wire, tape, or rope. Attach an insulator to each corner. Attach a ground-plane wire to each insulator. Bring the other ends of the ground-plane wires together, attach them to an insulator (C, Figure 7-13, page 7-18), and tie securely. Strip about 3 inches of insulation from each wire and twist them together.

(b) Tie one end of the radiating element wire to the other side of insulator C and the other end to another insulator (D, Figure 7-13). Strip about 3 inches of insulation from the radiating element at insulator C.

(c) Cut enough WD-1 field wire to reach from the proposed location of the antenna to the radio set. Keep this line as short as possible, because excess length reduces the efficiency of the system. Tie a knot at each end to identify it as the "hot" lead. Remove insulation from the "hot" wire and tie it to the radiating element wire at insulator C. Remove insulation from the other wire and attach it to the bare ground-plane element wires at insulator C. Tape all connections and do not allow the radiating element wire to touch the ground-plane wires.

(d) Attach a rope to the insulator on the free end of the radiating element and toss the rope over the branches of a tree. Pull the antenna as high as possible, keeping the lead-in routed down through the triangle. Secure the rope to hold the antenna in place.

(e) At the radio set, remove about 1 inch of insulation from the "hot" lead and about 3 inches of insulation from the other wire. Attach the "hot" line to the antenna terminal (doublet connector, if so labeled). Attach the other wire to the metal case-the handle, for example. Be sure both connections are tight or secure.

(f) Set up correct frequency, turn on the set, and proceed with communications.

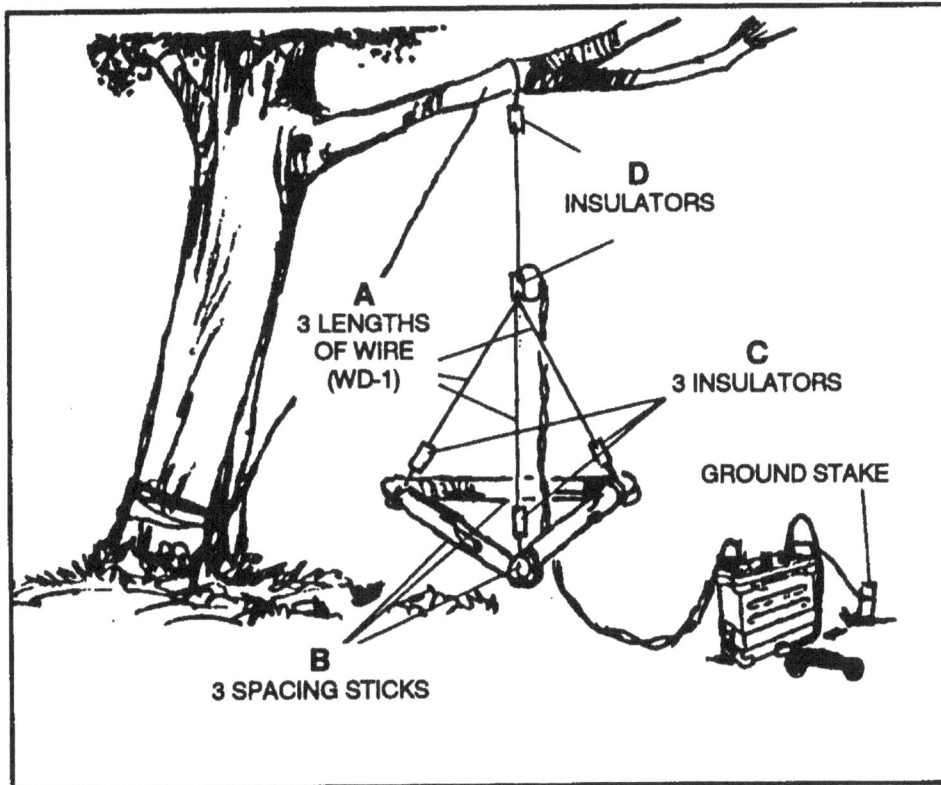

Figure 7-13. Expedient 292-type antenna.

(2) *Expedient patrol antenna. This* is another antenna that is easy to carry and quick to set up (Figure 7-14, page 7-20). The two radiating wires are cut to the length shown in Table 7-2 for the operating frequency. For the best results, the lead-in should extend at least 1.8 meters (6 feet) at right angles (plus or minus 30 degrees) to the antenna section before dropping to the radio set. The easiest way to set up this antenna is to measure the length of the radiating elements from one end of the lead-in (WD-1) and tie a knot at that point. The two wires are separated: one is lifted vertically by a rope and insulator;

the other is held down by a rock or other weight and a rope and insulator. The antenna should be as high as possible. The other end of the lead-in is attached to the radio set as described in paragraph 7-9c(l), expedient 292-type antenna.

OPERATING FREQUENCY IN MHz	ELEMENT LENGTH (radiating element and ground-plane elements)
30	2.38m (7 ft 10 in)
32	2.23m (7 ft 4 in)
34	2.1m (6 ft 11 in)
36	1.98m (6 ft 6 in)
38	1.87m (6 ft 2 in)
40	1.78m (5 ft 10 in)
43	1.66m (5 ft 5 in)
46	1.55m (5 ft 1 in)
49	1.46m (4 ft 9 in)
52	1.37m (4 ft 6 in)
55	1.3m (4 ft 3 in)
58	1.23m (4 ft 0 in)
61	1.17m (3 ft 10 in)
64	1.12m (3 ft 8 in)
68	1.05m (3 ft 5 in)
72	.99m (3 ft 3 in)
76	.94m (3 ft 1 in)

Table 7-2. Quick-reference chart.

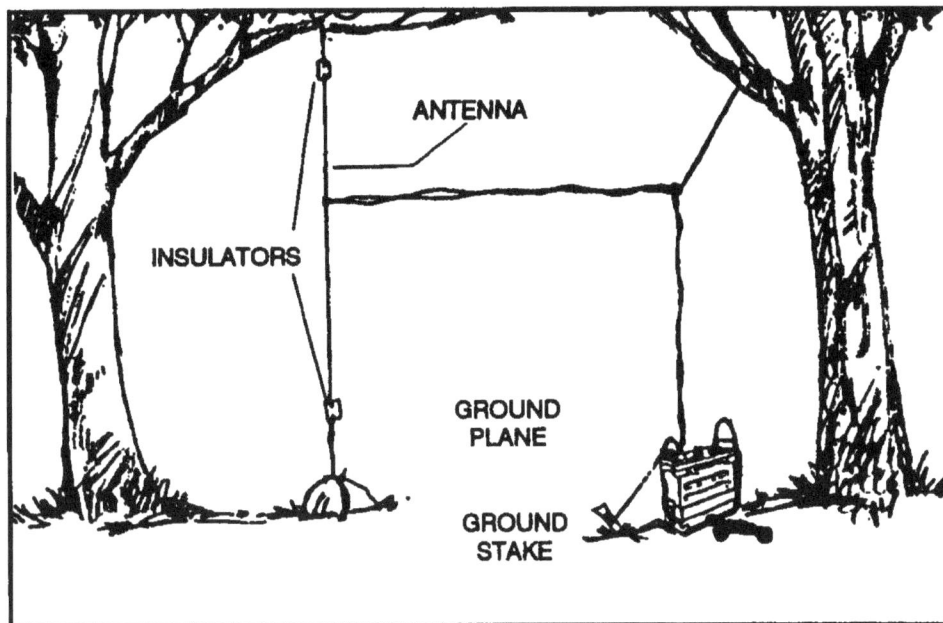

Figure 7-14. Expedient patrol antenna.

7-10. DESERT AREAS

Radio is usually the primary means of communications in the desert. It can be employed effectively in desert climate and terrain to provide a highly mobile means of communications demanded by widely dispersed forces.

a. **Techniques for Better Operations.** For the best operation in the desert, radio antennas should be located on the highest terrain available. In the desert, transmitters using whip antennas lose one-fifth to one-third of their normal range due to the poor electrical grounding common to desert terrain. For this reason, complete antenna systems must be used such as horizontal dipoles and vertical antennas with adequate counterpoises.

b. **Equipment Considerations.** Some radios automatically switch on their second blower fan if their internal temperature rises too high. Normally, this happens only in temperate climates when the radios are transmitting. This may disturb soldiers unaccustomed to radio operation in the desert environment. Operation of the second fan, however, is quite normal. Radio frequency power amplifiers used in AM and single sideband sets may overheat and burn out. Such equipment should be turned on only when necessary (signal reception is not affected). Since the RF power amplifiers take about 90 seconds to reach the operating mode, the SOP of units using the equipment allows for delays in replying. Dust affects communications equipment such as SSB/AMRF power amplifiers and radio teletypewriter sets. Radio teletypewriter sets are prone to damage due to the vulnerability of the oil lubrication system, which attracts and holds dust particles. Dust covers, therefore, should be used when possible. Some receiver-transmitter units have ventilating ports and channels that can get clogged with dust. These must be checked regularly and kept clean to prevent overheating.

c. **Batteries.** Dry battery supplies must be increased, since hot weather causes batteries to fail more rapidly.

d. **Electrical Insulation.** Wind-blown sand and grit damage electrical wire insulation over time. All cables that are likely to be damaged should be protected with tape before insulation becomes worn. Sand also finds its way into parts of items, such as "spaghetti cord" plugs, either preventing electrical contact or making it impossible to join the plugs together. A brush, such as an old toothbrush, should be carried and used to clean such items before they are joined.

e. **Condensation.** In deserts with relatively high dew levels and high humidity, overnight condensation can occur wherever surfaces are cooler than the air temperature, such as metals exposed to air. This condensation can affect electrical plugs, jacks, and connectors. All connectors likely to be affected by condensation should be taped to prevent moisture from contaminating the contacts. Plugs should be dried before inserting them into equipment jacks. Excessive moisture or dew should be dried from antenna connectors to prevent arcing.

f. **Static Electricity.** Static electricity is prevalent in the desert. It is caused by many factors, one of which is wind-blown dust particles. Extremely low humidity contributes to static discharges between charged particles. Poor grounding conditions aggravate the problem. All sharp edges (tips) of antennas should be taped to reduce wind-caused static discharges and the accompanying noise. If operating from a fixed position, teams ensure that equipment is properly grounded. Since static-caused noise lessens with an increase infrequency, the highest frequencies that are available and authorized should be used.

g. **Maintenance Improvement.** In desert areas, the maintenance of radio sets becomes more difficult due to the large amounts of sand, dust, or dirt that enter the equipment. Sets equipped with servomechanisms are especially affected. To reduce maintenance downtime, the team must keep sets in dustproof containers as much as possible. Air vent filters should also be kept clean to allow cool air to circulate to prevent overheating. Preventive maintenance checks should be made often. Also, the team should closely check the lubricated parts of the equipment. If dust and dirt mix with the lubricants, moving parts may be damaged.

7-11. MOUNTAINOUS AREAS

Operation of radios in mountainous areas have many of the same problems as in northern or cold weather areas. The mountainous terrain makes the selection of transmission sites a critical task In addition, terrain restrictions often require radio relay stations for good communications. Due to terrain obstacles, radio transmissions often have to be by line of sight. Also, the ground in mountainous areas is often a poor electrical conductor. Thus, a complete antenna system, such as a dipole or ground-plane antenna with a counterpoise, should be used. The maintenance procedures required in mountainous areas are the same as for northern or cold weather areas. The varied or seasonal temperature and climatic conditions in mountainous areas make flexible maintenance planning a necessity.

7-12. URBANIZED TERRAIN

Radio communications in urbanized terrain pose special problems. Some problems are similar to those encountered in mountainous areas. Some problems include obstacles blocking transmission paths, poor electrical conductivity due to pavement surfaces, and commercial power line interference.

a. Very high frequency radios are not as effective in urbanized terrain as they are in other areas. The power output and operating frequencies of these sets require a line of sight between antennas. Line of sight at street level is not always possible in built-up areas.

b. High frequency radios do not require or rely on line of sight as much as VHF radios. This is due to operating frequencies being lower and power output being greater. The problem is that HF radio sets are not organic to small units. To overcome this, the VHF signals must be retransmitted.

c. Retransmission stations in aerial platforms can provide the most effective means if available. Organic retransmission is more likely to be used. The antenna should be hidden or blended in with surroundings. This helps prevent the enemy from using it as a landmark to "home in" his artillery bombardment. Antennas can be concealed by water towers, existing civilian antennas, or steeples.

7-13. NUCLEAR BIOLOGICAL AND CHEMICAL ENVIRONMENT

One of the realities of fighting on today's battlefield is the presence of nuclear weapons. Most soldiers are aware of the effects of nuclear blast, heat, and radiation. The ionization of the atmosphere by a nuclear explosion will have degrading effects on communications due to static and the disruption of the ionosphere.

a. Electromagnetic pulse results from a nuclear explosion and presents a great danger to our radio communications. An EMP is a strong pulse of electromagnetic radiation, many times stronger than the static pulse generated by lightning. This pulse can enter the radio through the antenna system, power connections, and signal input connections. In the equipment, the pulse can break down circuit components such as transistors, diodes, and integrated circuits. It can melt capacitors, inductors, and transformers, destroying a radio.

b. Defensive measures against EMP call for proper maintenance, especially the shielding of equipment. When the equipment is not in use, all antennas and cables should be removed to decrease the effect of EMP on the equipment.

Section III
COMMUNICATIONS FORMATS

Timely, accurate information reporting reduces the unknown aspects of the enemy and the area of operations, contributing to the commander's risk assessment and successful application of combat power. This section provides the sniper team with a means of organized and rapid information delivery through reporting formats.

7-14. SPOT REPORT

This paragraph complies with STANAG 2022.

The sniper team uses the SPOTREP to report intelligence information. Each report normally describe a single observed event. When reporting groups of enemy vehicles, personnel report the location of the center of mass or indicate "from—to" coordinates. Higher headquarters sets the SPOTREP format, but the report usually follows the SALUTE format.

LINE 1 The size of the enemy force observed.

LINE 2 What the enemy was doing.

LINE 3 Where the enemy was located.

LINE 4 The unit to which the enemy belongs specified by markings on vehicles, distinctive features on uniforms, or special equipment that may identify the type enemy unit.

LINE 5 Time the enemy was observed.

LINE 6   Equipment the enemy carried, wore, or used.

Example: "C12, THIS IS STRIKER 1, SPOTREP, OVER."
"STRIKER 1, THIS IS C12 SEND MESSAGE, OVER."
"C12, THIS IS STRIKER 1. LINE 1: 3. LINE 2 MOVING IN A WESTERLY DIRECTION. LINE 3: GL024396. LINE 4: UNKNOWN. LINE 5: 2709911437. LINE 6: 1 SVD WITH PSO-1 TELESCOPE. CAMOUFLAGED OVERGARMENT AND RUCKSACK TWO INDIVIDUALS CARRYING AKM-74 RIFLES. 9-MM MAKAROV PISTOLS WITH SHOULDER HOLSTERS AND RUCKSACKS."

7-15. SITUATION REPORT

This paragraph complies with STANAG 2020.

The sniper team submits the SITREP to higher headquarters to report tactical situations and status. The team submits the report daily by 0600 hours after significant events or as otherwise required by the SEO or commander. The sender says, "SITREP," to alert the receiver of the type of report being sent. The following explains the reporting format according to line number:

LINE 1 Report as of date-time group.

LINE 2 Brief summary of enemy activity, casualties inflicted, prisoners captured.

LINE 3 Your location (encrypted—if not using secure communications).

LINE 4 Combat vehicles, operational.

    a. Improved TOW vehicle.

    b. M3 Bradley/ M113Al.

    c. M1.

    d. M60A3 tanks.

    e. M106A1 mortar carriers.

    f. Armored vehicle launched bridges (AVLB).

LINE 5 Defensive obstacles encoded.

    a. Coordinates of mine fields.

    b. Coordinates of demolitions executed.

    c. Coordinates of reserve demolition targets.

LINE 6 Personnel strength.

    a. Green (full strength, 90 percent or better on hand).

    b. Amber (reduced strength, 80 to 89 percent on hand).

    c. Red (reduced strength, 60 to 79 percent on hand, mission-capable).

    d. Black (reduced strength, 59 percent or lesson hand).

LINE 7   Class III and V for combat vehicles.

    a. Ammunition—green, amber, red, or black.

    b. Fuel—green, amber, red, or black.

LINE 8 Summary of tactical intentions.

Example: "RED 1, THIS IS RED 5; BLUE 2. LINE 1: 062230. LINE 2: NEGATIVE CONTACT. LINE 3: I SET ES, STA NEL. LINE 4B: 1. LINE 5: ABATIS, 1 SET XB, RDJ ALT. LINE 6: GREEN. LINE 7A: GREEN. LINE 7B: AMBER. LINE 8: CONTINUING MISSION."

7-16. RECONNAISSANCE REPORT

This paragraph complies with STANAG 2096.

Due to the length and detail of a reconnaissance report, it should be sent by messenger rather than transmitted by radio. Graphic overlays and sketches are normally included with the report. The following explains the reporting format according to line number:

LINE 1 OR HEADING (collection data).

    a. DTG information collected.

    b. DTG information received.

    c. Reporting unit.

LINE 2 OR 3 CAPITAL ROUTE CLASSIFICATION (data for a route classification).

    a. Start point.

    b. Checkpoint/ release point.

    c. Classification (code).

    d. Trafficability (code).

    e. Movement (code).

    f. Location of critical points.

LINE 3 OR BRIDGE CLASSIFICATION (data for a bridge classification).

    a. Location.

    b. One-way class.

    c. Two-way class.

    d. Overhead clearance.

    e. Bypass location.

    f. Bypass (code).

    g. Slope of entry bank.

    h. Slope of exit bank.

LINE 4 OR FORDING/ SWIM SITE (data for a ford or swim site).

    a. Location.

    b. Velocity (water speed).

    c. Depth.

    d. Type bottom (code).

    e. Width.

    f. Length.

    g. Slope of entry bank.

    h. Slope of exit bank.

LINE 5 OR TUNNEL CLASSIFICATION (data for a tunnel classification).

    a. Location.

    b. Usable width.

    c. Overhead clearance.

    d. Length.

    e. Bypass location.

LINE 6 OR OBSTACLES (obstacle information).

    a. Location.

    b. Slope (code).

    c. Type (code).

    d. Length.

    e. Bypass location.

f. Dimensions.

    (1) From:

    (2) To:

    (3) To:

CODES: Classification

    GREEN - all vehicles.

    AMBER - no AVLBs.

    RED - armed personnel carriers/ BFVs.

    BLACK -1 l/ 4-ton wheels or less.

Trafficability

    X - all weather.

    Y - limited weather.

    Z - fair weather.

Movement

    F - fast.

    S - slow.

Bypass

    E - easy.

    D - difficult.

Type bottom

    M - mud.

    C - clay.

    S - sand.

    G - gravel.

    R - rock.

    P - paving.

Slope

    A - less than 7 percent.

    B - 7 or 10 percent.

    C - 10 to 14 percent.

    D - Over 14 percent.

Type obstacle

   MF - mine field.

   TD - tank ditch.

   RF - rockfall or slide.

   CH - chemical.

   NBC - radiological.

   RB - roadblock.

   AB - abatis.

   O - other.

NOTES: 1. During reconnaissance., report items as they occur, since they are time-sensitive.

2. If time permits, submit overlays to the S2 during briefing. The S2 routinely consolidates details of terrain features and passes them to higher headquarters at the end of the debriefing.

Example: "C12, THIS IS STRIKER 1, RECONREP OVER."
"STRIKER 1, THIS IS Cl2; SEND MESSAGE, OVER."
 "C12, THIS IS STRIKER 1. LINE 1A: 2609910800. LINE lC: ST 1. LINE 2A: I SET DL, JAR CMN. LINE 2B: SIL MNC. LINE 2C: GREEN. LINE 2D: X. LINE 2E: F."

7-17. MEACONING, INTRUSION, JAMMING, AND INTERFERENCE REPORT.

This paragraph complies with STANAG 6004.

When the sniper team knows or suspects that the enemy is jamming, or knows or suspects that the enemy is intruding on the net, the incident is reported immediately by secure means to higher headquarter. Such information is vital for the protection and defense of friendly radio communications. The sniper who is experiencing the MIJI incident forwards this report through the chain of command to the unit OP. He also submits a separate report for each MIJI incident. An example of a MIJI 1 report follows:

   ITEM 1-022 (encrypted) or MIJI 1.

   ITEM 2-3 (encrypted) or JAMMING.

ITEM 3 - 1 (encypted) or RADIO.

ITEM 4 - 46.45 (encyypted if being transmitted over a nonsecure communications means).

ITEM 5 - N6B85S.

ITEM 6 - FA86345964 (encrypted if being transmitted over a nonsecure communications means).

a. Item 1 - Type of Report. When transmitted over nonsecure communications means, the numerals 022 are encrypted as Item 1 of the MIJI report. When transmitted over secure communications means, the term MIJI 1 is used as Item 1 of the MIJI 1 report.

b. Item 2 - Type of MIJI Incident. When transmitted over nonsecure communications means, the appropriate numeral preceding one of the items below is encrypted as Item 2 of the MIJI report. When transmitted over secure communications means, the appropriate term below is used as Item 2 of the MIJI 1 report.

- Meaconing.
- Intrusion.
- Jamming.
- Interference.

c. Item 3 - Type of Equipment Affected. When transmitted over nonsecure communications means, the appropriate numeral preceding one of the terms below is encrypted as Item 3 of the MIJI 1 report. When transmitted over secure communications means, the appropriate term below is used as Item 3 of the MIJI report.

- Radio.
- Radar.
- Navigational aid.
- Satellite.
- Electro-optics.

d. Item 4 - Frequency or Channel Affected. When transmitted over nonsecure communications means, the frequency or channel affected by the MIJI incident is encrypted as Item 4 of the MIJI 1 report. When transmitted over secure communications means, the frequency or channel affected by the MIJI incident is Item 4 of the MIJI 1 report.

e. Item 5 - Victim Designation and Call Sign of Affected Station Operator. The complete call sign of the affected station operator is Item 5 of the MIJI 1 report over both secure and nonsecure communications means.

f. Item 6 - Coordinates of the Affected Station. When transmitted over nonsecure communications means, the complete grid coordinates of the affected station are encrypted as Item 6 of the MIJI 1 report. When transmitted over secure communications means, the complete grid coordinates of the affected station are Item 6 of the MIJI 1 report.

7-18. SHELLING REPORTS

This paragraph complies with STANAG 2934.

The sniper team prepares and submits a SHELREP when it receives incoming rockets, mortars, or artillery rounds (FM 6-121). The team also uses this format for bombing attacks and mortars. The SHELREP format is as follows:

- ALPHA: Unit call sign.
- BRAVO: Location of observer.
- CHARLIE: Azimuth to flash or sound.
- DELTA: Time shelling started.
- ECHO: Time shelling ended.
- FOXTROT: Location of shelled area.
- GOLF: Number, type, and caliber (fire support team personnel only).
- HOTEL: Nature of fire (barrage, harassment, or registration).
- INDIA: Number of rounds.
- JULIET: Time of flash to bang.
- KILO: Damage.

7-19. ENEMY PRISONER OF WAR/CAPTURED MATERIEL REPORT

This paragraph complies with STANAG 2084.

The sniper team immediately tags EPWs and captured materiel. This ensures that information of intelligence value (place, time, and circumstances of capture) is not lost during evacuation. Only EPWs or materiel of immediate tactical importance are reported to the troop or battalion TOG Snipers use the following formats to report EPWs and captured materiel:

a. Enemy Prisoners of War.

LINE 1 - Type of report.

LINE 2 - Item captured.

LINE 3 - Date/ time of capture.

LINE 4 - Place of capture-grid coordinates.

LINE 5 - Capturing unit-all sign.

LINE 6 - Circumstances of capture (be brief).

b. Captured Materiel.

LINE 1 - Type of report.

LINE 2 - Item captured.

LINE 3 - Type document/ equipment.

LINE 4 - Date/ time captured.

LINE 5 - Place of capture-call sign.

LINE 6 - Capturing unit—call sign.

LINE 7 - Circumstances of capture (be brief).

After the report is given to the company/ team/ commander, disposition instructions will be provided if needed.

7-20. NBC 1 REPORT

This paragraph complies with STANAG 2103.

The sniper team uses the NBC 1 report to submit initial and subsequent information on an NBC attack, transmitting over the command or operation and intelligence net immediately after an NBC attack.

LINE 1 OR EVENT - Type of attack-nuclear, chemical, or biological.

LINE 2 OR BRAVO - Grid location of observer.

LINE 3 OR CHARLIE - Direction from observer to attack—mils or degree—true, grid, or magnetic.

LINE 4 OR DELTA - Date-time group of detonation or star of attack.

LINE 5 OR ECHO - Illumination time in seconds for nuclear attack.

LINE 6 OR ECHO BRAVO - End time for biological/ chemical attack

LINE 7 OR FOXTROT - Actual or estimated (state which) grid coordinates for location of attack.

LINE 8 OR GOLF - Means of delivery.

LINE 9 OR HOTEL - Height of nuclear burst in feet or meters and or type of burst.

LINE 10 OR HOTEL BRAVO - Type of biological/ chemical attack and height of burst.

LINE 11 OR INDIA BRAVO - Number of munitions or aircraft.

LINE 12 OR EFFECTS - Effects of burst/ agent on personnel.

LINE 13 OR JULIETT - Flash-to-bang time in seconds for nuclear attack.

LINE 14 OR KILO - Crater (yes or no) and width in meters.

LINE 15 OR KILO BRAVO - Vegetation chemical/ biological.

LINE 16 OR LIMA - Nuclear burst angular cloud width, measured at five minutes after detonation in mils or degrees.

LINE 17 OR MIKE - Stabilized cloud top height, in feet or meters, or angular cloud top angle, in degrees or mils, measured at H+10 minutes after detonation and stabilized cloud height, in feet or meters, or angular cloud bottom angle, in degrees or mils, measured at H+10.

LINE 18 OR PAPA ALPHA- Grid of predicted outline of external contours of hazardous cloud or area.

LINE 19 OR PAPA BRAVO - Downwind direction of nuclear cloud or duration of hazard in days.

LINE 20 OR SIERRA - Date-time group of reading for nuclear or detection time for biological/chemical.

LINE 21 OR YANKEE BRAVO - Effective downwind direction and wind speed.

LINE 22 OR ZULU ALPHA STABILITY - Air stability indicator.

LINE 23 OR ZULU ALPHA TEMPERATURE - Surface air temperature.

LINE 24 OR ZULU ALPHA HUMIDITY - Relative humidity range.

LINE 25 OR ZULU ALPHA WEATHER - Significant weather phenomena.

LINE 26 OR ZULU ALPHA COVER - Cloud cover.

LINE 27 OR NARRATIVE - Other significant observation.

LINE 28 - Not used.

LINE 29 OR AUTHENTICATION - Self-authentication, if required.

7-21. MEDICAL EVACUATION REQUEST

This paragraph complies with STANAG 3204.

The sniper team sends a MEDEVAC request to the medical team on the company command net.

a. When air assets are not available, the sniper team uses the ground evacuation format.

LINE 1 - Evacuation.

LINE 2 - Location for pickup (encode).

LINE 3 - Number of casualties.

LINE 4 - Category of patient(s).

A Urgent.

B Priority.

C Routine.

Use the letter of the appropriate subparagraph from Line 4 with the number of casualties in Line 3—for example, a2 means there are two urgent patients for evacuation.

b. When air assets are available, the sniper team uses the air evacuation format.

LINE 1 - Location.

LINE 2 - Radio frequency, call sign, and suffix.

LINE 3 - Precedence:

URGENT__ PRIORITY__ ROUTINE__ TACTICAL
IMMEDIATE__

LINE 4 - Special equipment.

LINE 5 - Number of patients by type:

Little__ Ambulator__

LINE 6 - Security of pickup site.

LINE 7 - Method of marking pickup size.

LINE 8 - Patient's nationality and status.

LINE 9 - NBC contamination.

c. The definitions of the categories of precedence follow:

(1) *Urgent.* Used for emergency cases for evacuation as soon as possible and no more than two hours to save life, limb, and eyesight.

(2) *Priority.* Used when the patient should be evacuated within four hours or his medical condition will deteriorate to an URGENT precedence.

(3) *Routine.* Requires evacuation, but the patient's condition is not expected to deteriorate within the next 24 hours.

(4) *Tactical immediate.* Used when the patient's condition is not urgent or priority, but evacuation is required as soon as possible so as not to endanger the requesting unit's tactical mission.

CHAPTER 8
TRACKING/COUNTERTRACKING

When a sniper follows a trail, he builds a picture of the enemy in his mind by asking himself questions: How many persons am I following? What is their state of training? How are they equipped? Are they healthy? What is their state of morale? Do they know they are being followed? To answer these questions, the sniper uses available indicators to track the enemy. The sniper looks for signs that reveal an action occurred at a specific time and place. For example, a footprint in soft sand is an excellent indicator, since a sniper can determine the specific time the person passed By comparing indicators, the sniper obtains answers to his questions. For example, a footprint and a waist-high scuff on a tree may indicate that an armed individual passed this way.

Section I
TRACKING

Any indicator the sniper discovers can be defined by one of six tracking concepts: displacement, stains, weather, litter, camouflage, and immediate-use intelligence.

8-1. DISPLACEMENT
Displacement takes place when anything is moved from its original position. A well-defined footprint or shoe print in soft, moist ground is a good example of displacement. By studying the footprint or shoe print, the sniper determines several important facts. For example, a print left by worn footgear or by bare feet may indicate lack of proper equipment. Displacement can also result from clearing a trail by breaking or cutting through heavy vegetation with a machete. These trails are obvious to the most inexperienced sniper who is tracking. Individuals may

unconsciously break more branches as they follow someone who is cutting the vegetation. Displacement indicators can also be made by persons carrying heavy loads who stop to rest; prints made by box edges can help to identify the load. When loads are set down at a rest halt or campsite, they usually crush grass and twigs. A reclining soldier also flattens the vegetation.

a. **Analyzing Footprints.** Footprints may indicate direction, rate of movement, number, sex, and whether the individual knows he is being tracked.

(1) If footprints are deep and the pace is long, rapid movement is apparent. Long strides and deep prints with toe prints deeper than heel prints indicate running (A, Figure 8-l).

(2) Prints that are deep, short, and widely spaced, with signs of scuffing or shuffling indicate the person is carrying a heavy load (B, Figure 8-l).

(3) If the party members realize they are being followed, they may try to hide their tracks. Persons walking backward (C, Figure 8-1) have a short, irregular stride. The prints have an unnaturally deep toe, and soil is displaced in the direction of movement.

(4) To determine the sex (D, Figure 8-l), the sniper should study the size and position of the footprints. Women tend to be pigeon-toed, while men walk with their feet straight ahead or pointed slightly to the outside. Prints left by women are usually smaller and the stride is usually shorter than prints left by men.

b. **Determining Key Prints.** The last individual in the file usually leaves the clearest footprints; these become the key prints. The sniper cuts a stick to match the length of the prints and notches it to indicate the width at the widest part of the sole. He can then study the angle of the key prints to the direction of march. The sniper looks for an identifying mark or feature, such as worn or frayed footwear, to help him identify the key prints. If the trail becomes vague, erased, or merges with another, the sniper can use his stick-measuring devices and, with close study, can identify the key prints. This method helps the sniper to stay on the trail. A technique used to count the total number of individuals being tracked is the box method. There are two methods the sniper can use to employ the box method.

(1) The most accurate is to use the stride as a unit of measure (Figure 8-2) when key prints can be determined. The sniper uses the set of key prints and the edges of the road or trail to box in an area to analyze. This method is accurate under the right conditions for counting up to 18 persons.

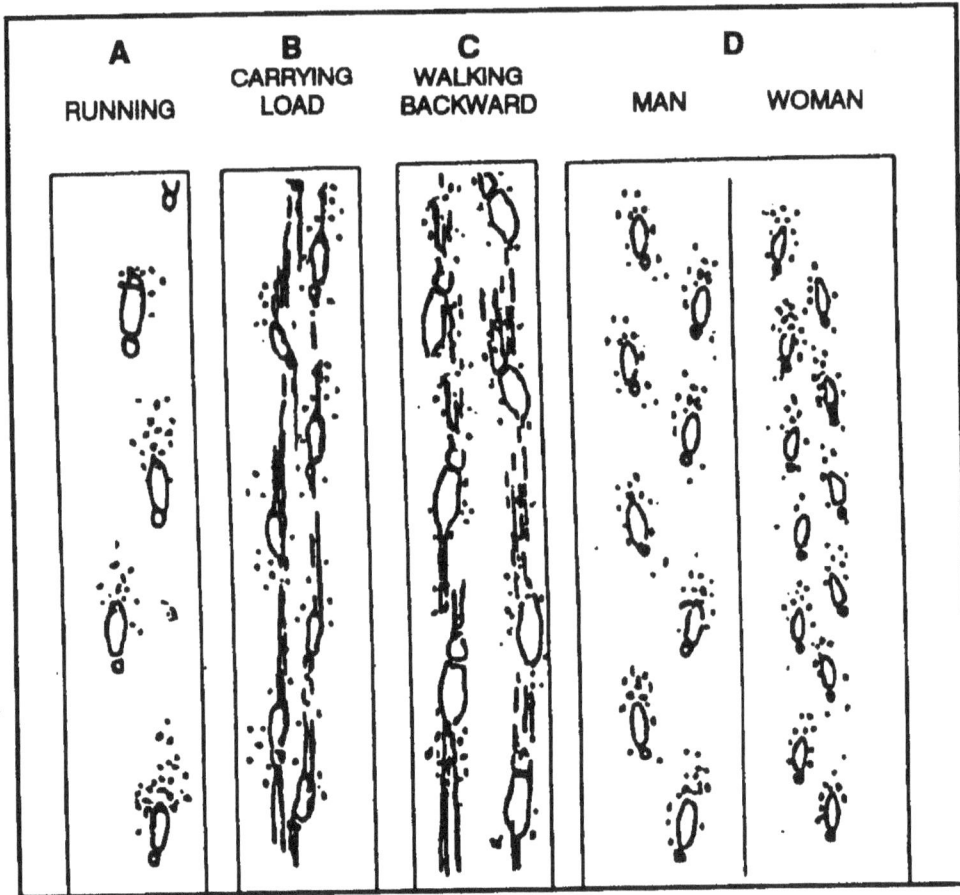

Figure 8-1. Different types of footprints.

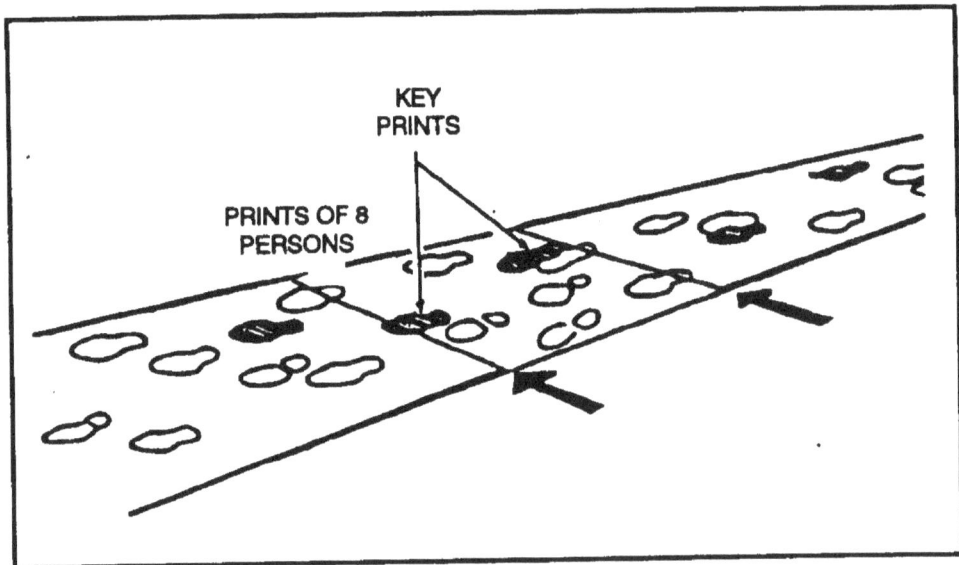

Figure 8-2. Stride measurement.

(2) The sniper may also use the the 36-inch box method (Figure 8-3) if key prints are not evident. To use the 36-inch box method, the sniper uses the edges of the road or trail as the sides of the box. He measures a cross section of the area 36 inches long, counting each indentation in the box and dividing by two. This method gives a close estimate of the number of individuals who made the prints; however, this system is not as accurate as the stride measurement.

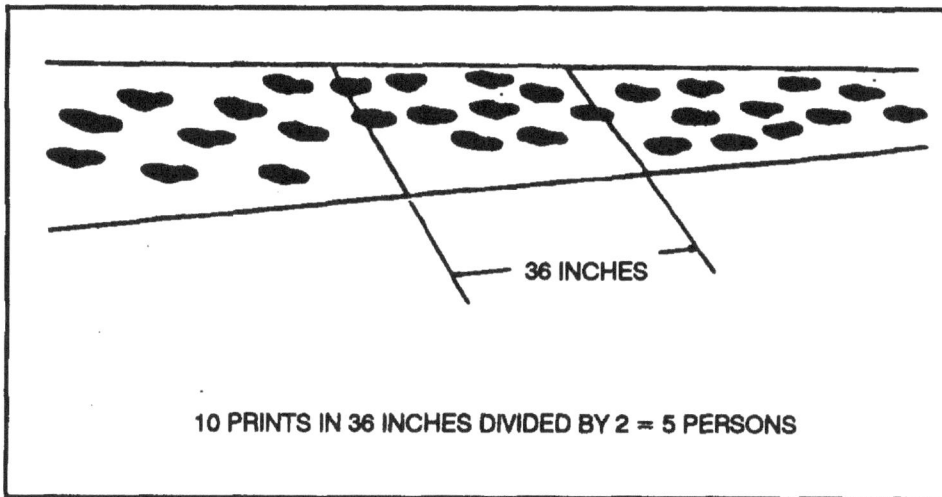

36 INCHES

10 PRINTS IN 36 INCHES DIVIDED BY 2 = 5 PERSONS

Figure 8-3. 36-inch box method.

c. **Recognizing Other Signs of Displacement**Foliage, moss, vines, sticks, or rocks that are scuffed or snagged from their original position form valuable indicators. Vines may be dragged, dew droplets displaced, or stones and sticks overturned (A, Figure 8-4) to show a different color underneath. Grass or other vegetation may be bent or broken in the direction of movement (B, Figure 8-4).

(1) The sniper inspects all areas for bits of clothing, threads, or dirt from footgear that can be torn or can fall and be left on thorns, snags, or the ground.

(2) Flushed from their natural habitat, wild animals and birds are another example of displacement. Cries of birds excited by unnatural movement is an indicator; moving tops of tall grass or brush on a windless day indicates that someone is moving the vegetation.

(3) Changes in the normal life of insects and spiders may indicate that someone has recently passed. Valuable clues are disturbed bees, ant holes uncovered by someone moving over them, or tom spider webs. Spiders often spin webs across open areas, trails, or roads to trap flying insects. If the tracked person does not avoid these webs, he leaves an indicator to an observant sniper.

(4) If the person being followed tries to use a stream to cover his trail, the sniper can still follow successfully. Algae and other water plants can be displaced by lost footing or by careless walking. Rocks can be displaced from their original position or overturned to indicate a lighter or darker color on the opposite side. The person entering or exiting a stream creates slide marks or footprints, or scuffs the bark on roots or sticks (C, Figure 8-4). Normally, a person or animal seeks the path of least resistance; therefore, when searching the stream for an indication of departures, snipers will find signs in open areas along the banks.

A
TURNED OVER ROCKS AND STICKS

B
CRUSHED AND DISTURBED

C
SLIP MARK AND WATER FILLED FOOTPRINTS ON STREAM BANKS

Figure 8-4. Other displacements.

8-2. STAINS

A stain occurs when any substance from one organism or article is smeared or deposited on something else. The best example of staining is blood from a profusely bleeding wound. Bloodstains often appear as spatters or drops and are not always on the ground; they also appear smeared on leaves or twigs of trees and bushes.

a. By studying bloodstains, the sniper can determine the wound's location.

(1) If the blood seems to be dripping steadily, it probably came from a wound on the trunk.

(2) If the blood appears to be slung toward the front, rear, or sides, the wound is probably in the extremity.

(3) Arterial wounds appear to pour blood at regular intervals as if poured from a pitcher. If the wound is veinous, the blood pours steadily.

(4) A lung wound deposits pink, bubbly, and frothy bloodstains.

(5) A bloodstain from a head wound appears heavy, wet, and slimy.

(6) Abdominal wounds often mix blood with digestive juices so the deposit has an odor and is light in color.

The sniper can also determine the seriousness of the wound and how far the wounded person can move unassisted. This proms may lead the sniper to enemy bodies or indicate where they have been carried.

b. Staining can also occur when muddy footgear is dragged over grass, stones, and shrubs. Thus, staining and displacement combine to indicate movement and direction. Crushed leaves may stain rocky ground that is too hard to show footprints. Roots, stones, and vines may be stained where leaves or berries are crushed by moving feet.

c. The sniper may have difficulty in determining the difference between staining and displacement since both terms can be applied to some indicators. For example, muddied water may indicate recent movement; displaced mud also stains the water. Muddy footgear can stain stones in streams, and algae can be displaced from stones in streams and can stain other stones or the bank. Muddy water collects in new footprints in swampy ground; however, the mud settles and the water clears with time. The sniper can use this information to indicate time; normally, the mud clears in about one hour, although time varies with the terrain.

8-3. WEATHER

Weather either aids or hinders the sniper. It also affects indicators in certain ways so that the sniper can determine their relative ages. However, wind, snow, rain, or sunlight can erase indicators entirely and hinder the sniper. The sniper should know how weather affects soil, vegetation, and other indicators in his area. He cannot determine the age of indicators until he understands the effects that weather has on trail signs.

a. By studying weather effects on indicators, the sniper can determine the age of the sign (for example, when bloodstains are fresh, they are bright red). Air and sunlight first change blood to a deep ruby-red color, then to a dark brown crust when the moisture evaporates. Scuff marks on trees or bushes darken with time; sap oozes, then hardens when it makes contact with the air.

b. Weather affects footprints (Figure 8-5). By carefully studying the weather process, the sniper can estimate the age of the print. If particles of soil are beginning to fall into the print, the sniper should become a stalker. If the edges of the print are dried and crusty, the prints are probably about one hour old. This varies with terrain and should be considered as a guide only.

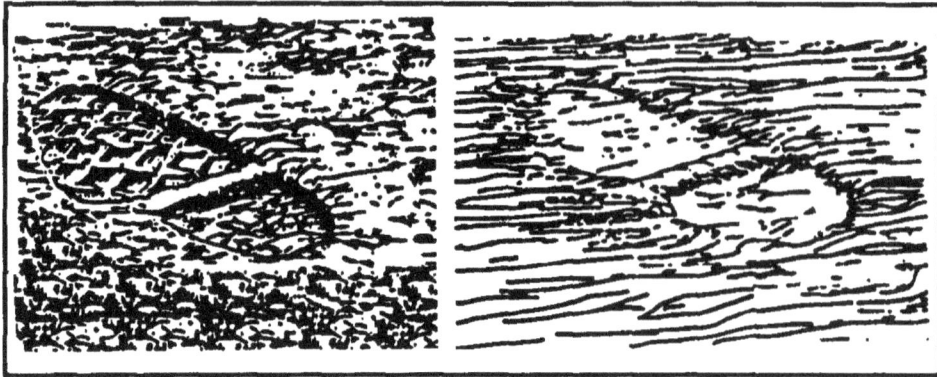

Figure 8-5. Weather effects on footprints.

c. A light rain may round the edges of the print. By remembering when the last rain occurred, the sniper can place the print into a time frame. A heavy rain may erase all signs.

d. Trails exiting streams may appear weathered by rain due to water running from clothing or equipment into the tracks. This is especially true if the party exits the stream single file. Then, each person deposits water into the tracks. The existence of a wet, weathered trail slowly fading into a dry trail indicates the trail is fresh.

e. Wind dries tracks and blows litter, sticks, or leaves into prints. By recalling wind activity, the sniper may estimate the age of the tracks. For example, the sniper may reason "the wind is calm at the present but blew hard about an hour ago. These tracks have litter in them, so they must be over an hour old." However, he must be sure that the litter was not crushed into them when the prints were made.

(1) Wind affects sounds and odors. If the wind is blowing toward the sniper, sounds and odors may be carried to him; conversely, if the wind is blowing *away* from the sniper, he must be extremely cautious since wind also carries sounds toward the enemy. The sniper can determine wind direction by dropping a handful of dust or dried grass from shoulder height. By pointing in the same direction the wind is blowing, the sniper can localize sounds by cupping his hands behind his ears and turning slowly. When sounds are loudest, the sniper is facing the origin.

(2) In calm weather (no wind), air currents that may be too light to detect can carry sounds to the sniper. Air cools in the evening and moves downhill toward the valleys. If the sniper is moving uphill late in the day or at night, air currents will probably be moving toward him if no other wind is blowing. As the morning sun warms the air in the valleys, it moves uphill. The sniper considers these factors when plotting patrol

routes or other operations. If he keeps the wind in his face, sounds and odors will be carried to him from his objective or from the party being tracked.

(3) The sun should also be considered by the sniper. It is difficult to fire directly into the sun, but if the sniper has the sun at his back and the wind in his face, he has a slight advantage.

8-4. LITTER
A poorly trained or poorly disciplined unit moving over terrain may leave a trail of litter. Unmistakable signs of recent movement are gum or candy wrappers, food cans, cigarette butts, remains of fires, or human feces. Rain flattens or washes litter away and turns paper into pulp. Exposure to weather can cause food cans to rust at the opened edge; then, the rust moves toward the center. The sniper must consider weather conditions when estimating the age of litter. He can use the last rain or strong wind as the basis for a time frame.

8-5. CAMOUFLAGE
Camouflage applies to tracking when the followed party employs techniques to baffle or slow the sniper. For example, walking backward to leave confusing prints, brushing out trails, and moving over rocky ground or through streams.

8-6. IMMEDIATE-USE INTELLIGENCE
The sniper combines all indicators and interprets what he has seen to form a composite picture for on-the-spot intelligence. For example, indicators may show contact is imminent and require extreme stealth.

a. The sniper avoids reporting his interpretations as facts. He reports what he has seen rather than stating these things exist. There are many ways a sniper can interpret the sex and size of the party, the load, and the type of equipment. Timeframes can be determined by weathering effects on indicators.

b. Immediate-use intelligence is information about the enemy that can be used to gain surprise, to keep him off balance, or to keep him from escaping the area entirely. The commander may have many sources of intelligence reports, documents, or prisoners of war. These sources can be combined to form indicators of the enemy's last location, future plans, and destination.

c. Tracking, however, gives the commander definite information on which to act immediately. For example, a unit may report there are no men of military age in a village. This information is of value only if it is combined with other information to make a composite enemy picture in the area. Therefore, a sniper who interprets trail signs and reports that he is 30 minutes behind a known enemy unit, moving north, and located at a specific location, gives the commander information on which he can act at once.

8-7. DOG/HANDLER TRACKING TEAMS

Dog/ handler tracking teams are a threat to the sniper team. While small and lightly armed, they can increase the area that a rear area security unit can search. Due to the dog/ handler tracking team's effectiveness and its lack of firepower, a sniper team may be tempted to destroy such an "easy" target. Whether a sniper should fight or run depends on the situation and the sniper. Eliminating or injuring the dog/ handler tracking team only confirms that there is a hostile team operating in the area.

a. When looking for sniper teams, trackers use wood line sweeps and area searches. A wood line sweep consists of walking the dog upwind of a suspected wood line or brush line. If the wind is blowing through the woods and out of the wood line, trackers move 50 to 100 meters inside a wooded area to sweep the wood's edge. Since wood line sweeps tend to be less specific, trackers perform them faster. An area search is used when a team's location is specific such as a small wooded area or block of houses. The search area is cordoned off, if possible, and the dog/ handler tracking teams are brought on line, about 25 to 150 meters apart, depending on terrain and visibility. The handler trackers then advance, each moving their dogs through a specific corridor. The handler tracker controls the dog entirely with voice commands and gestures. He remains undercover, directing the dog in a search pattern or to a likely target area. The search line moves forward with each dog dashing back and forth in assigned sectors.

b. While dog/ handler tracking teams area potent threat, there are counters available to the sniper team. The beat defenses are basic infantry techniques: good camouflage and light, noise, and trash discipline. Dogs find a sniper team either by detecting a trail or by a point source such as human waste odors at the hide site. It is critical to try to obscure or limit trails around the hide, especially along the wood line or area closest to the team's target area. Surveillance targets are usually the major axis of advance. "Trolling the wood lines" along likely looking roads or intersections is a favorite tactic of dog/ handler tracking teams. When moving into a target area, the sniper team should take the following countermeasures:

(1) Remain as faraway from the target area as the situation allows.

(2) Never establish a position at the edge of cover and concealment nearest the target area

(3) Reduce the track. Try to approach the position area on hard, dry ground or along a stream or river.

(4) Urinate in a hole and cover it up. Never urinate in the same spot.

(5) Bury fecal matter deep. If the duration of the mission permits, use MRE bags sealed with tape and take it with you.

(6) Never smoke.

(7) Carry all trash until it can be buried elsewhere.

(8) Surround the hide site with a 3-cm to 5-cm band of motor oil to mask odor; although less effective but easier to carry, garlic may be used. A dead animal can also be used to mask smell, although it may attract unwanted canine attention.

c. If a dog/ handler tracking team moves into the area, the sniper team can employ several actions but should first check wind direction and speed. If the sniper team is downwind of the estimated search area, the chances are minimal that the team's point smells will probably be detected. If upwind of the search area, the sniper team should attempt to move downwind. Terrain and visibility dictate whether the sniper team can move without being detected visually by the handlers of the tracking team. Remember, sweeps are not always conducted just outside of a wood line. Wind direction determines whether the sweep will be parallel to the outside or 50 to 100 meters inside the wood line.

(1) The sniper team has options if caught inside the search area of a line search. The handlers rely on radio communications and often do not have visual contact with each other. If the sniper team has been generally localized through enemy radio detection-finding equipment, the search net will still be loose during the initial sweep. A sniper team has a small chance of hiding and escaping detection in deep brush or in woodpiles. Larger groups will almost certainly be found. Yet, the sniper team may have the opportunity to eliminate the handler and to escape the search net.

(2) The handler hides behind cover with the dog. He searches for movement and then sends the dog out in a straight line toward the front. Usually, when the dog has moved about 50 to 75 meters, the handler calls the dog back. The handler then moves slowly forward and always from covered position to covered position. Commands are by voice and gesture with a backup whistle to signal the dog to return. If a handler is eliminated or badly injured after he has released the dog, but before he has recalled it, the dog continues to randomly search out and away from the handler. The dog usually returns to another handler or to his former

handler's last position within several minutes. This creates a gap from 25 to 150 meters wide in the search pattern. Response times by the other searchers tend to be fast. Given the high degree of radio communication, the injured handler will probably be quickly missed from the radio net. Killing the dog before the handler will probably delay discovery only by moments. Dogs are so reliable that if the dog does not return immediately, the handler knows something is wrong.

(3) If the sniper does not have a firearm, one dog can be dealt with relatively easy if a knife or large club is available. The sniper must keep low and strike upward using the wrist, never overhand. Dogs are quick and will try to strike the groin or legs. Most attack dogs are trained to go for the groin or throat. If alone and faced with two or more dogs, the sniper should avoid the situation.

Section II
COUNTERTRACKING

If an enemy tracker finds the tracks of two men, this may indicate that a highly trained team may be operating in the area. However, a knowledge of countertracking enables the sniper team to survive by remaining undetected.

8-8. EVASION

Evasion of the tracker or pursuit team is a difficult task that requires the use of immediate-action drills to counter the threat. A sniper team skilled in tracking techniques can successfully employ deception drills to lessen signs that the enemy can use against them. However, it is very difficult for a person, especially a group, to move across any area without leaving signs noticeable to the trained eye.

8-9. CAMOUFLAGE

The sniper team may use the most used and the least used routes to cover its movement. It also loses travel time when trying to camouflage the trail.

a. **Most Used Routes.** Movement on lightly traveled sandy or soft trails is easily tracked. However, a sniper may try to confuse the tracker by moving on hard-surfaced, often-traveled roads or by merging with civilians. These routes should be carefully examined; if a well-defined approach leads to the enemy, it will probably be mined, ambushed, or covered by snipers.

b. **Least Used Routes.** Least used routes avoid all man-made trails or roads and confuse the tracker. These routes are normally magnetic azimuths between two points. However the tracker can use the proper concepts to follow the sniper team if he is experienced and persistent.

c. **Reduction of Trail Signs.** A sniper who tries to hide his trail moves at reduced speed· therefore the experienced tracker gains time. Common methods to reduce trail signs areas follows:

(1) Wrap footgear with rags or wear soft-soled sneakers which make footprints rounded and leas distinctive.

(2) Brush out the trail This is rarely done without leaving signs.

(3) Change into footgear with a different tread immediately following a deceptive maneu er.

(4) Walk on hard or rocky ground.

8-10. DECEPTION TECHNIQUES

Evading a skilled and persistent enemy tracker requires skillfully executed maneuvers to deceive the tracker and to cause him to lose the trail. An enemy tracker cannot be outrun by a sniper team that is carrying equipment because he travels light and is escorted by enemy forces designed for pursuit. The size of the pursuing force dictates the sniper team s chances of success in employing ambush-type maneuvers. Sniper teams use some of the following techniques in immediate-action drills and deception drills.

a. **Backward Walking.** One of the basic techniques used is that of walking backward (Figure 8-6) in tracks already made and then stepping off the trail onto terrain or objects that leave little sign. Skillful use of this maneuver causes the tracker to look in the wrong direction once he has lost the trail.

b. **Large Tree** A good deception tactic is to change directions at large trees (Figure 8-7). To do this the sniper moves in any given direction and walks past a large tree (12 inches wide or larger) from 5 to 10 paces. He carefully walks backward to the forward side of the tree and makes a 90-degree change in the direction of travel passing the tree on its forward side. This technique uses the tree as a screen to hide the new trail from the pursuing tracker.

> NOTE: By studying signs a tracker may determine if an attempt is being made to confuse him. If the sniper team loses the tracker by walking backward footprints will be deepened at the toe and soil will be scuffed or dragged in the direction of movement. By following carefully the tracker can normally find turnaround point.

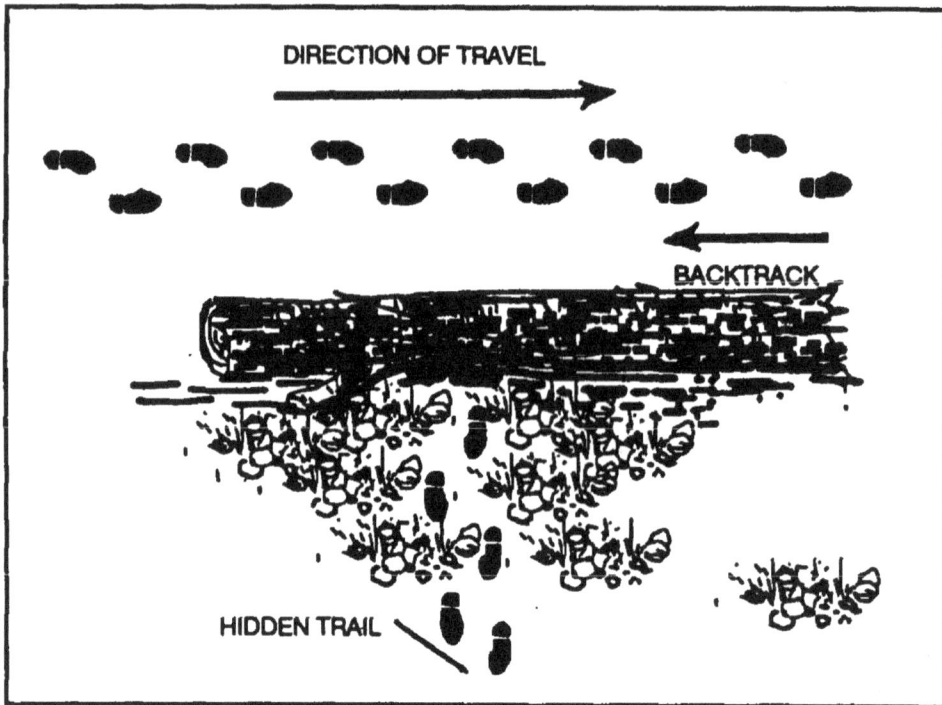

Figure 8-6. Walking backward.

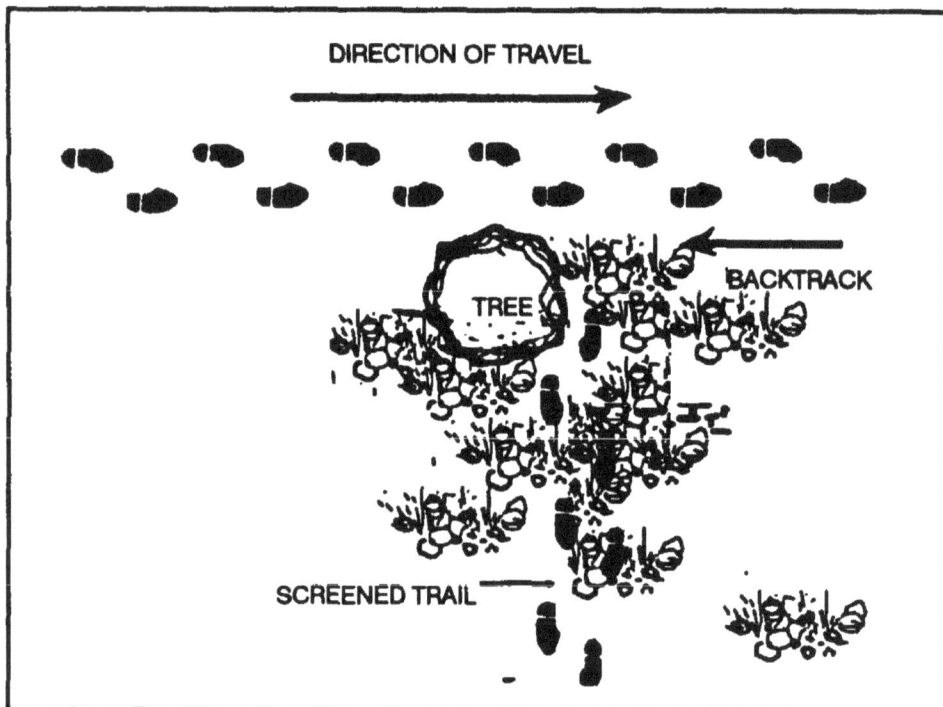

Figure 8-7. Large tree.

c. **Cut the Corner.** Cut-the-corner technique is used when approaching a known road or trail. About 100 meters from the road, the sniper team changes its direction of movement, either 45 degrees left or right. Once the road is reached, the sniper team leaves a visible trail in the same direction of the deception for a short distance on the road. The tracker should believe that the sniper team "cut the corner" to save time. The sniper team backtracks on the trail to the point where it entered the road, and then it carefully moves on the road without leaving a good trail. Once the desired distance is achieved, the sniper team changes direction and continues movement (Figure 8-8).

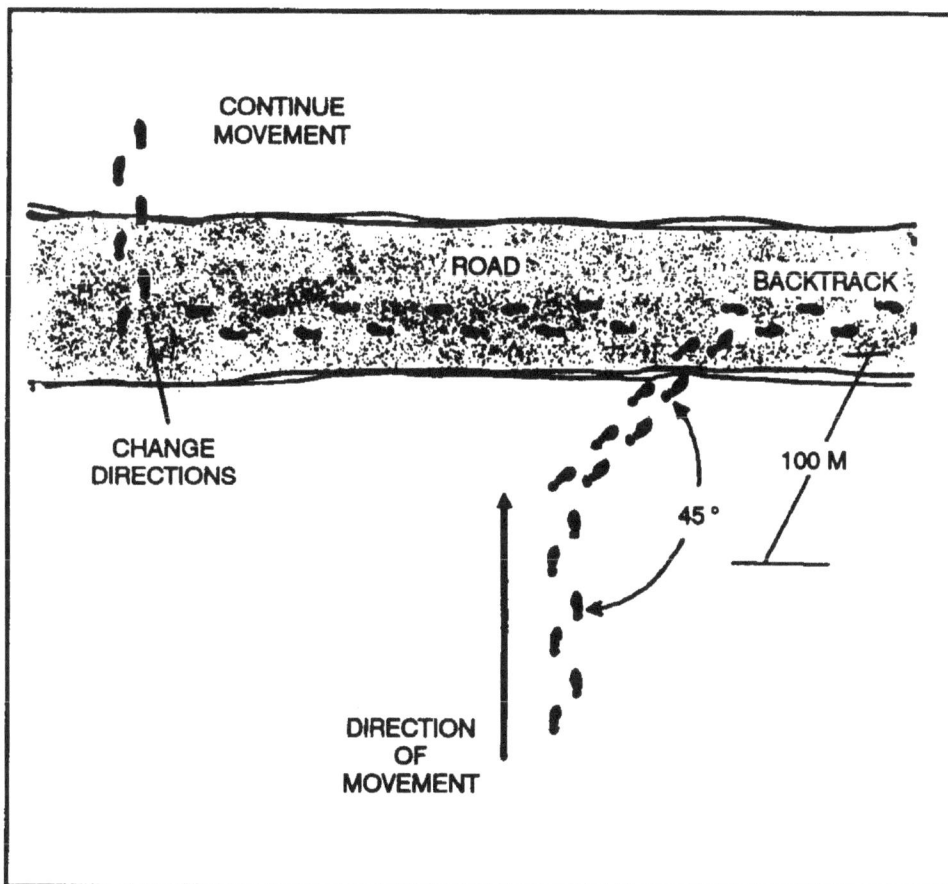

Figure 8-8. Cut the corner.

d. **Slip the Stream.** The sniper team uses slip-the-stream technique when approaching a known stream. The sniper team executes this method the same as the cut the corner technique. The sniper team establishes the 45-degree deception maneuver upstream, then enters the stream. The sniper team moves upstream to prevent floating debris and silt from compromising its direction of travel, and the sniper team establishes false trails upstream if time permits. Then, it moves downstream to escape since creeks and streams gain tributaries that offer more escape alternatives (Figure 8-9).

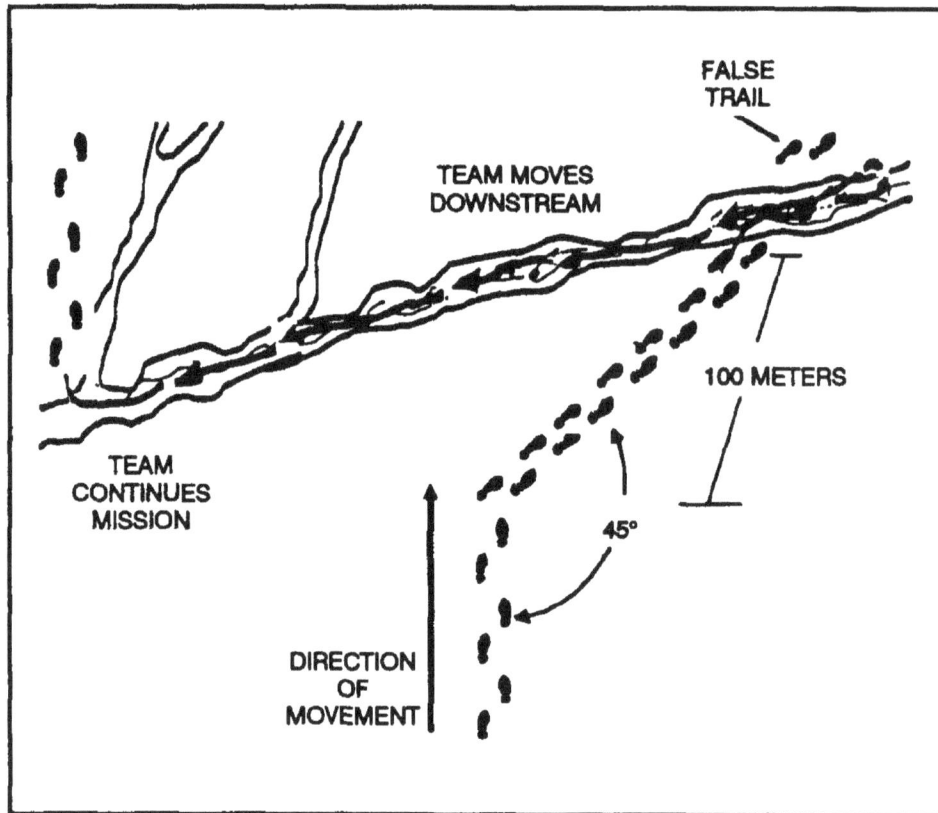

Figure 8-9. Slip the stream.

e. **Arctic Circle.** The sniper team uses the arctic circle technique in snow-covered terrain to escape pursuers or to hide a patrol base. It establishes a trail in a circle (Figure 8-10, page 8-16) as large as possible. The trail that starts on a road and returns to the same start point is effective. At some point along the circular trail, the sniper team removes snowshoes (if used) and carefully steps off the trail, leaving one set of tracks. The large tree maneuver can be used to screen the trail. From the hide position, the sniper team returns over the same steps and carefully fills them with snow one at a time. This technique is especially effective if it is snowing.

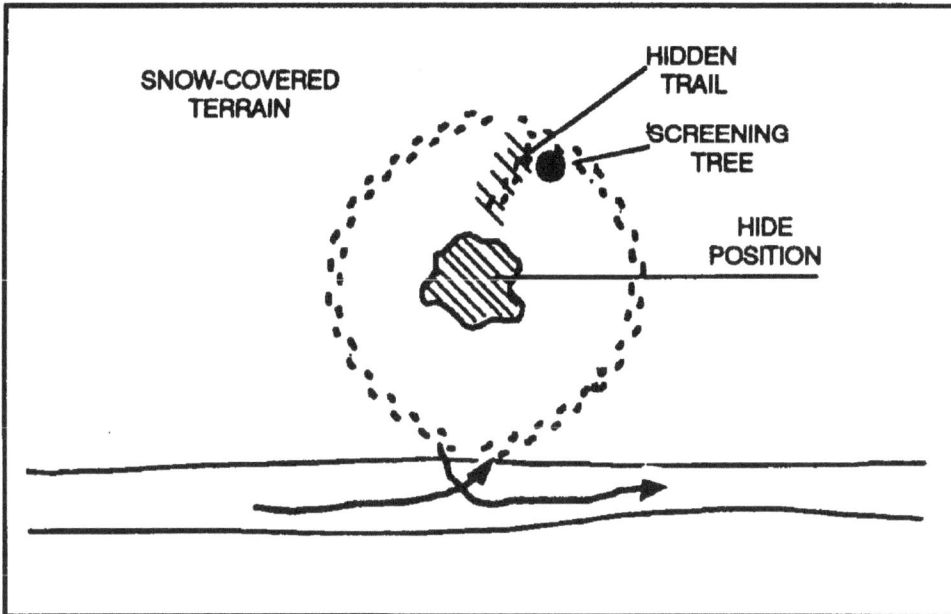

Figure 8-10. Arctic circle.

f. **Fishhook.** The sniper team uses the fishhook technique to double back (Figure 8-11) on its own trail in an overwatch position. The sniper team can observe the back trail for trackers or ambush pursuers. If the pursuing force is too large to be destroyed, the sniper team strives to eliminate the tracker. The sniper team uses the hit-and-run tactics, then moves to another ambush position. The terrain must be used to advantage.

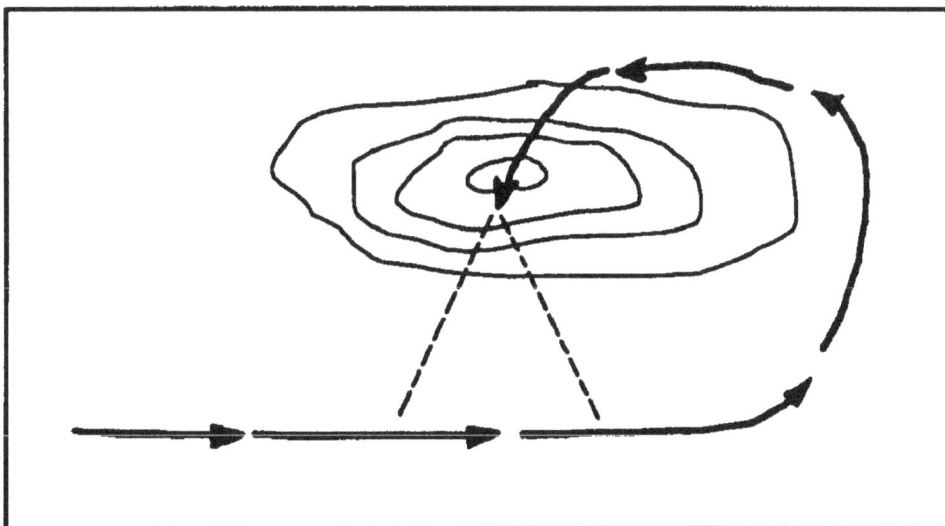

Figure 8-11. Fishhook.

CHAPTER 9
SNIPER SUSTAINMENT TRAINING

Repetitive training in long-range markmanship and field-craft skills ensures the best probability of effective engagement and the minimum risk of detection. Snipers must sustain basic soldier skills and master and sustain critical mission skills to accomplish their objectives. Both sniper and observer are trained snipers and should be highly skilled in the art of sniping. Sniping skills perish quickly; therefore, sniper teams must sustain and sharpen those skills regularly. To deny the importance and need to sustain sniper training deprives the commander of a valuable asset. This chapter also includes a 5-day sniper sustainment training program.

9-1. BASIC SKILLS SUSTAINMENT

Due to the primary and secondary missions of the sniper, minimum skill sustainment should include observation, range estimation, concealment, concealed movement, and rifle firing. Sustainment of these skills may best be accomplished through sniper training exercises and unit-level live-fire exercises. (DA Pamphlet 350-38 outlines the frequency and ammunition requirements needed to conduct sniper training.) Sniper training exercises provide snipers with practical experience in detecting and engaging realistic targets under field conditions on ranges comparable to a battlefield. This training also provides snipers with a means to practice the various sniper training fundamentals that has been taught previously, often collectively. These exercises mayor may not be graded; however, competition is a proven method to obtain the desired results. At the end of the exercises, the trainer critiques each sniper on his performance. These exercises include zeroing and practice fire, field fire (unknown distance), concealment, concealed movement target detection, range estimation, land navigation, memory enhancement exercise (KIM game), and communications. Each sniper will go through these training exercises.

a. **Zeroing and Pratice Fire.** To engage targets effectively during training exercises and in combat, the sniper must have his rifle accurately zeroed. For this reson the zeroing exercises are normally conducted on a measured known-distance range to ensure precise adjustment, recording, and practice under ideal conditions and to eliminate variables that may prevent achieving an effective zero. The sniper rifle is zeroed using both the telescopic and iron sights. A bull's-eye-type target should be used for zeroing. It is important to acquire a point-of-aim, point-of-impact zero at 100 meters using the M24. As the distance increases, the sniper must adjust his telescope to allow for elevation and wind to ensure the rounds stay in the center of the target.

b. **Field Fire.** Practical firing exercises are designed to develop sniper proficiency in the accurate and rapid engagement of various combat-type targets, as well as to provide practical work in other field techniques. Snipers should be given positions on the firing line and areas of the field fire course to observe and make range cards of the area.

(1) After the range cards have been completed, the snipers will be required to fire the course by having one member call the wind and adjust the other member's fire. The ability to call the wind is important as successful engagement of the targets. After one member fires the course, they switch positions and repeat the fire course.

(2) When firing the course, snipers should engage the targets in a sequence that starts with the 200-meter target, then engage each target out to 800 meters, then engage targets back to the 200-meter target. (Targets are engaged twice. Snipers will engage a target with no more than two rounds per target.) The course consists of engaging 20 targets with 30 rounds of ammunition within a 30-minute time iimit. The sniper should be scored as follows:

- 10 points for first-round hits.
- 5 points for second-round hits.
- 200 points maximum.
- 140 points needed to pass (70 percent).

(3) To enhance training, snipers should also fire the field fire course during limited visibility with overhead illumination such as parachute flares. This puts stress on the sniper to determine the range and to engage a target in a short amount of time.

(4) lb provide the most realistic training environment trainers do not use range commands to commence fire and cease fire in sniper exercises.

The only exception to this is when an unsafe condition exists. The command CEASE FIRE should be given immediately. Snipers must be given a thorough orientation on each exercise (to include safety requirements) before they are permitted to move into position. After the sniper has assumed his firing position in the designated location, he should be allowed to fire without further commands. Therefore, the range must be cleared for firing before the exercise begins. An NCO (assistant trainer) must be with each sniper to keep score and to maintain safety during the exercise. When the sniper completes firing, the NCO ensures the rifle is clear and signals the range officer.

NOTE: A blank copy of the forms that follow are located at the back of this manual for local reproduction.

c. **Concealment.** Concealment exercises develop and test the sniper's ability to conceal himself in an expedient firing position while observing and engaging an observer-instructor. Figure 9-1, page 9-4, is an example of completed DA Form 7325-R, Concealment Exercise Scorecard.

(1) In a cleared area with a wood line about 100 meters away, snipers conceal themselves within 10 minutes in the wood line. After the 10-minute preparation, an observer-instructor 100 meters away visually searches the area for 2 minutes without the aid of optics. After 2 minutes, the observer-instructor searches the wood line (from his position) for 18 minutes, using binoculars and the M49 observation telescope. If there are more than 10 snipers in the exercise, two observer-instructors and two assistant trainers may be needed. After the 20-minute period, an assistant trainer with a radio moves within 10 feet of a sniper, who is ready to fire at an observer-instructor.

(2) The sniper should be able to identify a white 5-inch number that is painted on an 8-inch by 8-inch international orange panel. This panel is held over a vital part of the observer-instructor, and two blanks are fired at him without the sniper being detected. If the target detects the sniper, he radios the assistant trainer and directs him to the sniper. The exercise should be scored on a 10-point system, with 7 points being a passing score. (See Paragraph 9-4, Day 3, to score the concealment exercise.)

CONCEALMENT EXERCISE SCORECARD
Exercise Number ___

For use of this form, see FM 23-10. The poponent agency is TRADOC.

DATA REQUIRED BY PRIVACY ACT OF 1974.

AUTHORITY: 10 USC 3012(g)/Executive Order 1974. PRINCIPAL PURPOSE(S): Evaluates individual training. ROUTINE USE(S): Evaluates individual proficiency. SSN is used for positive identification purposes only. MANDATORY OR VOLUNTARY DISCLOSURE AND EFFECT ON INDIVIDUAL NOT PROVIDING INFORMATION: Voluntary. Individuals not providing information cannot be rated/scored on a mass basis.

Last name	First	MI	Rank	SSN	Unit
SMITH	JOHN	D	SFC	457-28-1738	Dco 2-29 IN

Date	Weather\visibility	Score
17 FEB 93	CLEAR 67° UNLIMITED	9

	Points	Deducted	Total
If the sniper			
• Was detected without the aid of optics (first 2 minutes)	2	0	2
• Was detected with the aid of optics (18 minutes)	1	0	3
• Was detected when assistant trainer was within 10 feet of sniper	1	0	4
• Properly identied the number within 30 seconds	1	0	5
• Failed to properly identify the number	0	3	2
• Fired first shot, not detected	4	0	6
• Fired second shot, not detected	1	0	7
• Maintained stable firing position (support)	2	0	9
• Properly adjusted weapon's scope for range and windage	1	0	10

(Check one of the target indicators.)

☐ Contrast to background
☐ Muzzle blast
☑ Muzzle flash
☐ Improper movement techniques
☐ Improper camouflage
☐ shine
☐ Outline
☐ Sound

John D. Smith
Sniper's signature

NOTES:

1. If the sniper was caught trying to identify the number, score 4 points.

2. If muzzle blast/flash is detected, deduct 1 point from total score.

3. Failing to comply with training standards and objectives (such as unnecessary movement, premature fire, outside prescribed boundries) will result in termination of the exercise and a score of zero.

Roy B. Jones
Trainer's signature

DA FORM 7325-R, JUL 94

Figure 9-1. Example of completed DA Form 7325-R, Concealment Exercise Scorecard.

d. **Concealed Movement.** Concealed movement exereise develops and tests the sniper's ability to move and occupy a firing position undetected. Trainers record scores on DA Form 7326-R, Concealed Movement Exercise Scorecard (Figure 9-2, page 9-6).

(1) This exercise requires the same amount of trainers and equipment as in the concealment exercises. Areas used should be observable for 1,000 meters and have easily recognizable left and right limits. Ideally, snipers should train in a different type of area each time they perform these exercises.

(2) The snipers move 800 to 600 meters toward two observer-instructors, occupy a firing position 100 to 200 meters away, identify in the same manner as the concealment exercise, and fire two blanks at the targets without being detected at any time. If one of the observer-instructors detects a sniper, he radios one of the assistant trainers and directs him to the sniper's position. The sniper is given three hours to complete the exercise. The exercise is scored on a lo-point system, with 7 points being a passing wore. (See Paragraph 9-4, Day 4, to score concealed movement exercise.)

CONCEALED MOVEMENT EXERCISE SCORECARD

Exercise Number _____

For use of this form, see FM 23-10. The proponent agency is TRADOC.

DATA REQUIRED BY PRIVACY ACT OF 1974.

AUTHORITY: 10 USC 3012(g)/Executive Order 9397. PRINCIPAL PURPOSE(S): Evaluates individual training. ROUTINE USE(S): Evaluates individual proficiency. SSN is used for positive identification purposes only. MANDATORY OR VOLUNTARY DISCLOSURE AND EFFECT ON INDIVIDUAL NOT PROVIDING INFORMATION: voluntary. Individuals not providing information cannot be rated/scored on a mass basis.

Last name	First	MI	Rank	SSN	Unit
SMITH	JOHN	D	SFC	457-28-0739	Dco 2-29 IN

Date	Weather\visibility	Score
17 FEB 93	CLEAR 50' UNLIMITED	

	Points	Deducted	Total
If the sniper			
•Was detected moving to FFL	0	0	0
•Was detected moving in FFL	6	0	6
•Fired first round shot, not detected	2	0	8
•Was not detected when assistant trainer is within 10 feet of sniper	2	0	10
•Properly identified number (within 30 seconds) .	2	0	12
•Failed to properly identify number	2	0	14
•Was not detected when assistant trainer is within 5 feet of sniper	2	0	16
•Fired second shot, not detected	2	0	18
•Maintained stable firing position (support) . .	1	0	19
•Properly adjusted weapon's scope for range and windage	1	0	20

(Check one of the target indicators.)

☐ Contrast to back-ground | ☐ Improper camouflage
■ Muzzle blast | ☐ Shine
☐ Muzzle flash | ☐ Outline
☐ Improper movement techniques | ☐ sound

NOTES:
1. If muzzle blast/flash is detected, deduct 1 point from total score.
2. Failing to comply with training standards and objectives (such as unnecessary movement, premature fire, outside prescribed boundries) will result in termination of the exercise and a score of zero.

REMARKS: Explain in detail on back the reason for sniper's detection.

Roy B Wood
Trainer's signature

John D. Smith
Sniper's signature

DA FORM 7326-R, JUL 94

Figure 9-2. Example of completed DA Form 7326-R, Concealed Movement Exercise Scorecard.

e. **Target Detection.** Target detection exercises sharpen the sniper's eyes by requiring him to detect, describe, and plot objects that cannot be easily seen or described without the skillful use of optics. Scores are recorded on DA Form 7327-R, Target Detection Exercise Scorecard (Figure 9-3, page 9-8).

(1) Areas used for target detection should be partly cleared at least 200 meters in depth and 100 meters in width with easily definable left and right limits. The area should have at least three TRPs that are easily recognized and positioned in different locations throughout the area. Ten military items are placed in the area. These items can be radio antennas, small-scale mock vehicles, batteries, map protractors, or weapons. Items should be placed so that they are undetectable with the naked eye, detectable but indescribable with the binoculars, and describable only by using the M49 observation telescope.

(2) Snipers are given an M49 observations telescope, M22 binoculars, pencil, clip board, and scorecard. Snipers are given 40 minutes to detect, describe, and plot each item in the area. Snipers remain in the prone position throughout the exercise. After 15 minutes, they will move to a different position, left or right of the centerline of observation and remain there for the next 15 minutes. For the last 10 minutes, they can choose a position anywhere along the line. When an object is detected, the sniper gives his location on the line of observation (A or B). Next, the sniper must describe the object using the categories of size, shape, color, condition, and appearance. Snipers receive 1/2 point for correctly plotting a target and 1/2 point for correctly describing it. They must achieve 7 points to receive a GO in this area.

NOTE: The trainer should sanitize the site before the exercise. If the sniper finds additional items to describe he may use the eleventh and twelfth lines of the scorecard. If the trainer allows the sniper can obtain credit for observation and detection skills.

Figure 9-3. Example of completed DA Form 7327-R, Target Detection Exercise Scorecard.

The form in the figure contains:

ROSTER #: (11) TARGET DETECTION EXERCISE SCORECARD EX #: (PE.2)
For use of this form, see FM 23-10; the proponent agency is TRADOC

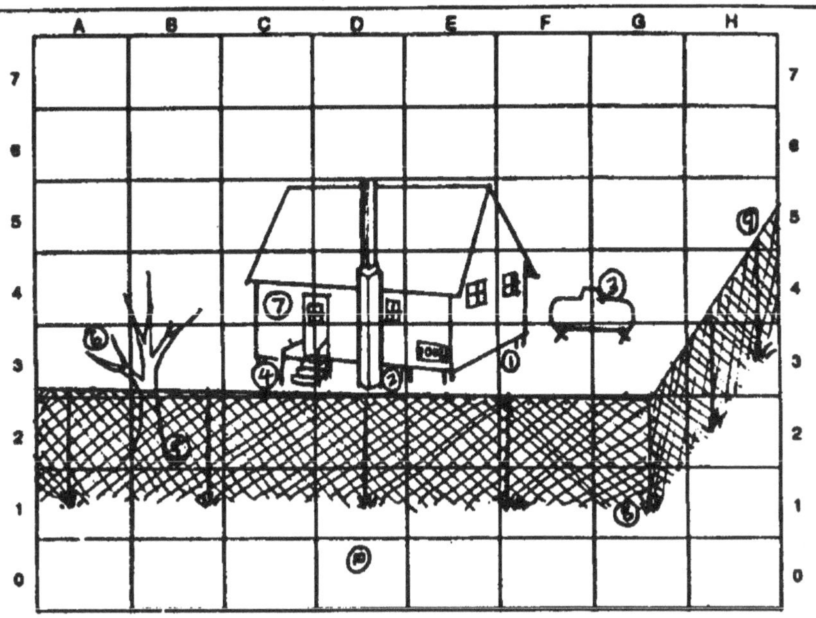

SKETCH NAME: BLDG 1002
GRID COORDINATE: PL217003
WEATHER: CLEAR 72°
Magnetic Azimuth: .75
Sketch # 1 of 1
Block Scale: 1" = 10M
NAME: EDISON
RANK: SGT
DATE: 1 APR 93 TIME: 1650

#	SIZE	SHAPE	COLOR	CONDITION	APPEARS TO BE	GRID BOX LOC.
1	36"×4"×2"		BLACK	SERV	M16A2	F-3 Ⓐ
2	10"×1"×4"		GREEN	SERV	RADIO	D-3 Ⓐ
3	3"×3"×3"		BLUE	UNSERV	GRENADE	G-4 Ⓐ
4	5"×2"×6"		BLACK	SERV	TANK	C-3 Ⓑ
5	4"×4"×1/64"		CLEAR	SERV	PROTRACTOR	B-2 Ⓑ
6	1"×1/8"×36"		YELLOW	SERV	STRAP	A-3 Ⓑ
7	1/32"×1/2"×1/4"		BLACK	SERV	RANK INSIGNIA	C-3 Ⓐ
8	6"×4"×12"		GREEN	SERV	AMMO BOX	G-1 Ⓑ
9	1"×1"×1"		BLACK	SERV	CANTEEN CAP	H-4 Ⓐ
10	1/2"×1/2"×1"		SILVER	SERV	PAPER CLIP	D-0 Ⓑ

DA FORM 7327-R, JUL 94

f. **Range Estimation.** Snipers must correctly estimate distance to effectively fire weapons, complete accurate range cards, and give reliable intelligence reports. Range estimation exercises should be conducted in an area that allows unobstructed observation of a human-size target up to 1,000 meters away. Scores are recorded on DA Form 7328-R, Range Estimation Exercise Scorecard (Figure 9-4, page 9-10). Personnel should be placed at various ranges and stages of concealment to give the sniper a challenging and realistic exercise. Snipers should be graded on their ability to estimate range by using the naked eye, M19/ M22 binoculars, and the M3A scope. Snipers must correctly estimate the distance to 7 of 10 objects using their eyes (± 15 percent), 7 to 10 objects using the binoculars (± 10 percent), and 7 to 10 objects using the M3A telescope (± 5 percent). They must sketch their assigned sector on the back of the form, page 9-11.

RANGE ESTIMATION EXERCISE SCORECARD

Exercise Number___

For use of this form, see FM 23-10. The proponent agency is TRADOC.

DATA REQUIRED BY PRIVACY ACT OF 1974.

AUTHORITY: 10 USC 3012(g)/Executive Order 9397. PRINCIPAL PURPOSE(S): Evaluates individual training. ROUTINE USE(S): Evaluates individual proficiency. SSN is used for positive identification purposes only. MANDATORY OR VOLUNTARY DISCLOSURE AND EFFECT ON INDIVIDUAL NOT PROVIDING INFORMATION: Voluntary. Individuals not providing information cannot be rated/scored on a mass basis.

Last name	First	MI	Rank	SSN	Unit
SMITH	JOHN	D	SFC	457-28-038	D 6-2-29 IN

Date	Weather/visibility		Score
17 FEB 93	CLEAR 40' UNLIMITED		27

EYE ESTIMATION +- 15%	BINOCULAR ESTIMATION +- 10%	M3A TELESCOPE ESTIMATION +- 5%
1 100 M	1 105 M	1 108 M
2 130	2 X 142	2 127
3 200	3 217	3 215
4 225	4 232	4 X 215
5 215	5 260	5 258
6 300	6 315	6 315
7 350	7 330	7 326
8 400	8 410	8 405
9 X 500	9 420	9 440
10 475	10 405	10 488

Ray B. Qwle
Trainer's signature

DA FORM 7328-R, JUL 94

1. Within three minutes, the range to the target is estimated at each point, using the naked eye, binoculars, and the M3A telescope. Estimations must be performed in the order listed.

2. Once an estimate is recorded, it cannot be changed; it will be counted as incorrect. However, the M3A telescope estimate may be changed before the next set of estimates are recorded.

3. The use of calculators is encouraged.

4. This is an individual exercise. Any sniper that talks or tries to look at another sniper's scorecard is terminated from the exercise.

5. If there are any questions, the trainer will assist you.

John D. Smith
Sniper's signature

Figure 9-4. Example of completed DA Form 7328-R, Range Estimation Exercise Scorecard (front).

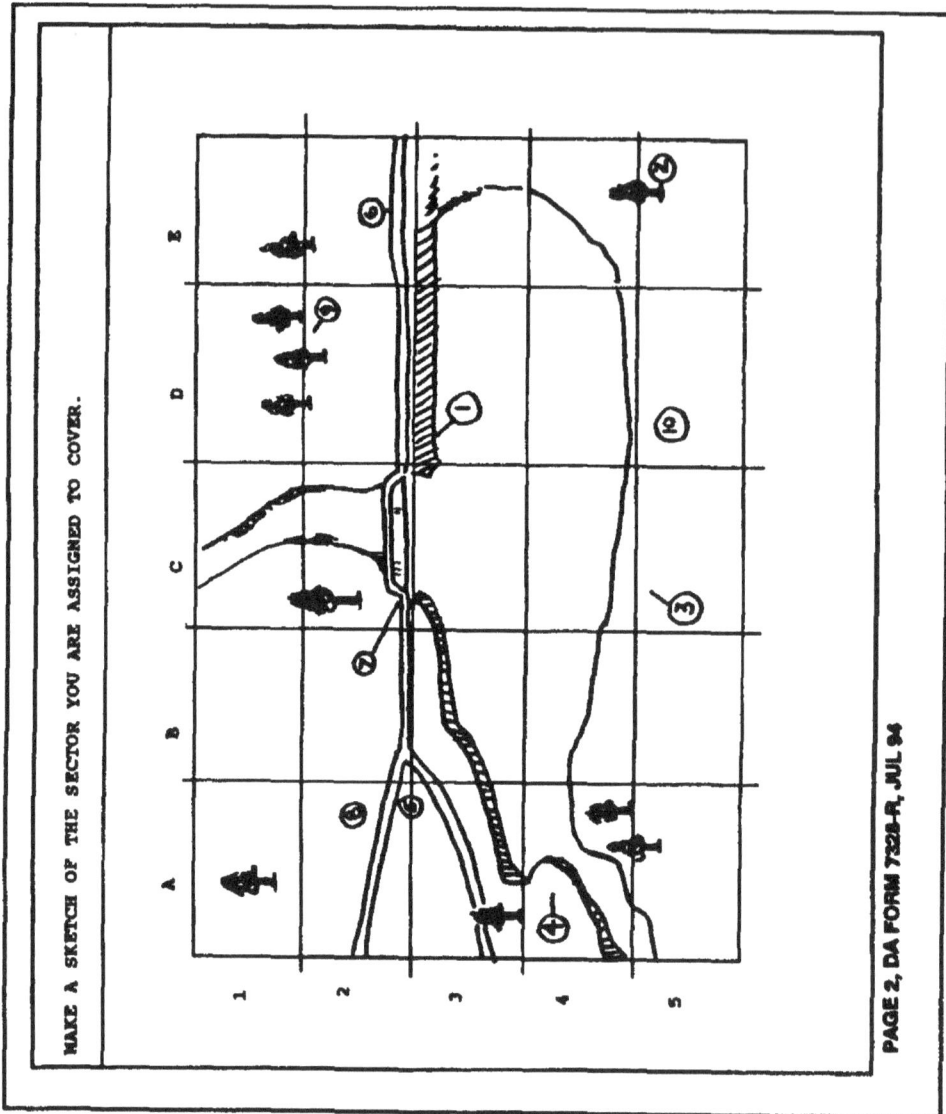

Figure 9-4. Example of completed DA Form 7328-R, Range Estimation Exercise Scorecard (back) (continued).

g. **Land Navigation.** This exercise develops the snipers' proficiency in specific field techniques such as movement, land navigation, and radiotelephone procedure. Snipers must move from a starting point to a specific location and then report. During this exercise, snipers should be fully equipped. (See Chapter 2.) To provide training under varied conditions, this exercise should be conducted at least twice, once during daylight and once during limited visibility.

(1) This exercise can beheld at the same time as the firing exercises. Half of the training class or group could conduct the land navigation exercise, while the other half conducts the firing exercise. When they finish, they change over.

(2) Snipers are assembled at the starting point and instructed on the mission objective, the observation positions, and the radio call signs. Trainers conduct an equipment check and an exercise briefing. This exercise requires snipers to move from the starting point to the designated location in less than two hours. They are instructed to avoid the observation positions, which represent the enemy. They must report their location every 15 minutes and their arrival at the destination site. A team starts the exercise with 100 points. The following point deductions are made for errors:

(a) Take 1 point off for each minute over the authorized two hours.

(b) Take 3 points off for every 5 meters that the sniper misses the designated destination.

(c) Take 5 points off for each instance of improper radio procedure or reporting.

(d) Take 10 points off for each time the sniper is seen by someone in the observation positions.

(e) Take 100 points off for being lost and failing to complete the exercise.

(3) At the end of this exercise, the trainer critiques the snipers' performance.

h. **Memory Enhancement Exercise (KIM Game).** A KIM game exercise consists of 10 variable military items on a table, covered with a blanket poncho, or anything suitable. Snipers observe the objects when uncovered but cannot touch the items or talk during the exercise. (Figure 9-5 is an example of a locally fabricated KIM game exercise scoresheet format.)

(1) After a prescribed time, the items are covered, and the snipers write their observations on a score sheet. They write the details that accurately describe the object, omitting unnecessary words. There are many variations that can be incorporated into a KIM game, such as PT, an extended amount of time between observing and recording, distractions

while observing and recording, or the use of different methods to display items. For example instead of a blanket uses towel or slides. At the end of the time limit, snipers turn in the score sheets, and trainers identify each item. Snipers describe each object in the following categories:

(a) *Size:* The sniper describes the object by giving the rough dimensions in a known unit of measure or in relation to a known object.

(b) *Shape:* The sniper describes the object by giving the shape such as round, square, or oblong.

(c) *Color:* The sniper records the color of the object.

(d) *Condition:* The sniper describes the object by giving the general or unusual condition of the object such as new, worn, or dented.

(e) *Appears to be:* The sniper describes what the object appears to be such as an AK-47 round or radio handset.

KIM GAME EXERCISE

NAME: BAILEY, WILLIAM   DATE 7 APR 93

ROS # Ø6

TEAM # A

KIMS GAME # G-1

SCORE _____

	SIZE	SHAPE	COLOR	CONDITION	APPEARS TO BE
1	1"×8"×1"		BLACK/GRAY	SERV	STAPLER
2	1/2"×2"×1/2"		GOLD	SERV	MHB-SPECIAL BALL
3	4"×8"×5"		CAMO	SERV	BDU CAP
4	10"×3"×9"		BLACK	SERV	JUNGLE BOOT
5	2"×12"×4"		BLACK	SERV	PVS-4
6	1/8"×1"×1/8"		BLACK	UNSERV	E-5 PIN ON
7	1/4"×1"×1"		TAN	SERV	EARPLUG CASE
8	2"×2"×6"		GREEN	UNSERV	COMPASS
9	3"×2"×8"		GREEN	SERV	AMMO POUCH
10	3"×1/4"×50"		GREEN	UNSERV	PISTOL BELT

Figure 9-5. Example of suggested format for KIM game exercise score sheet.

(2) Snipers receive 1/2 point for indicating that there was an item with some sort of description and the other 1/2 point for either exactly naming the item or giving a sufficiently detailed description using the categories listed above. The description must satisify the trainer to the extent that the sniper had never seen the object before. The total possible score is 10 points. Experience in the exercise, time restraints, and complexity of the exercise determines a passing score. This is the trainer's judgment based on his own experience in KIM games (Figure 9-6). The first few games should be strictly graded, emphasizing details. When the snipers are familiar with the game pattern, the trainer may make changes. The last game of the training should be identical to the first. In this way, the sniper can see if he improved.

KIM GAME SCHEDULE			
NO.	OBSERVE (minutes)	RECORD (minutes)	REMARKS
1	2:00	3:00	NO DISTRACTIONS
2	2:00	3:00	NOISE DURING RECORDING
3	1:50	2:50	FIRE BLANK WHILE RECORDING
4	1:50	2:50	PT BETWEEN OBSERVE/RECORD
5	1:30	2:30	2-HOUR DELAY BETWEEN OBSERVE/RECORD
6	REPEAT GAME NO. 1		

Figure 9-6. Example of suggested KIM game schedule.

i. **Communications.** Snipers must be highly trained in using the SOI and proper communication procedures. Maintaining communication is a primary factor in mission success. Areas of emphasis should include the following:
- Operation and maintenance of radios.
- Entering the net.
- Authentication.

- Encoding/ decoding.
- Encrypting/ decrypting.
- Antenna repair.
- Field-expedient antennas.

9-2. ADDITIONAL SKILLS SUSTAINMENT

Other than basic skills, the trainer must include additional skills into the sniper sustainment training program. Once mastered, these skills enhance the sniper's chance of surviving and accomplishing the mission.

a. **Call for Fire.** With advanced camouflage and movement techniques, snipers can move about the battlefield undetected. Snipers that have a working knowledge in the use and application of artillery, NGF, and CAS will bean asset to the commander. (See FM 6-30.)

(1) *Artillery fire.* Artillery fire is the secondary weapon of the sniper. Each sniper should master call-for-fire procedures (Figure 9-7, page 9-16), target location methods (Figure 9-8, page 9-17), and indirect-weapon system capabilities (Table 9-1, page 9-19). Separate radio stations may beset up with one being a simulated FDC. After the FDC receives the call for fire, it determines how the target will be attacked. That decision is announced to the FO as a message to the observer, which consists of three elements as follows:

- Unit to fire for effect.
- Any changes to requests in the call for fire.
- Method of fire (number of rounds to be fired).

Snipers can simulate calls for fire using the example format in Figure 9-7, page 9-16.

(2) *Naval gunfire and close air support.* In today's battlefield of "high-tech" munitions and delivery systems, a working knowledge of acquiring NGF and CAS (helicopter and fixed-wing) enables snipers to inflict heavy damage on enemy forces.

1. OBSERVER IDENTIFICATION

2. WARNING ORDER

   a. Type of Mission.
      (1) Adjust fire.
      (2) Fire for effect.
      (3) Suppress.

   b. Size of Element to Fire.
   c. Method of Target Location.
      (1) Grid.
      (2) Polar.
  (3) Shift from a known point.

3. TARGET LOCATION

   a. Grid. Six-digit grid coordinates.
   b. Polar. Direction and distance from the FO to the target.
   c. Shift. Direction to the target.
      (1) Lateral shift   L/R
      (2) Range shift   ±
      (3) Vertical shift   U/D

4. TARGET DESCRIPTION

5. METHOD OF ENGAGEMENT

   a. Type of Adjustment.
      (1) Area fire.*
         • Bracketing.*
         • Creeping (DANGER CLOSE).
      (2) Precision fire.
   b. Trajectory.*
   c. Ammunition.
      (1) Projectile.*
      (2) Fuze.*
      (3) Volume of fire.*
   d. Distribution.*

Figure 9-7. Call-for-fire-format.

6. METHOD OF FIRE AND CONTROL

  a. Method of Fire.
    (1) Center platoon/center section.
    (2) Battery/platoon right (left).
    (3) Time interval.

  b. Method of Control.
    (1) Fire when ready.*
    (2) At my command.
    (3) Cannot observe.
    (4) Time on target.
 *Standard

Figure 9-7. Call-for-fire-format (continued).

1. GRID (a) Determine a six-digit grid coordinate to the designated target.

       (b) Determine the grid direction (observer-target) to the target, and ensure that the O-T direction is sent to the FDC after the call for fire is completed before the first correction.

2. POLAR (a) Determine the O-T direction to the target from the FO's position.

       (b) Determine the distance from the FO's position to the target.

3. SHIFT (a) Determine the O-T direction to the target.

       (b) Determine the lateral shift from the known point to the target.

          $W = R \times M$ (mil relation).

          $W =$ Width of lateral shift in meters.

Figure 9-8. Target location methods.

R = Distance to the known point divided by 1,000. When shifting from a known point, the R is rounded to the nearest tenth.

M = Measured angle in mils between the known point and the target.

Example: $\dfrac{2,800}{1,000}$ (distance to known point) = 2/8 = R

M = 130 mils (measured angle from the known point to the target).

Therefore: W = R (2.8) × M (130)

W = 364 or LEFT 360 (nearest 10 mils)

(c) Determine the range shift from the known point to the target.

Example: 2,800 (distance to the known point)
          -1,700 (distance to the target)

           1,100 meters or DROP 1,100 meters
                 (nearest 100 meters)

Figure 9-8. Target location methods (continued).

WEAPON	MAXIMUM RANGE (METERS)	MINIMUM RANGE (METERS)	MAXIMUM RATE (ROUNDS PER MINUTE FIRST MINUTE)	SUSTAINED RATE (ROUNDS PER MINUTE)
FIELD ARTILLERY				
105-mm HOWITZER M101A1, TOWED	11,000 14,500 (RAP)	0	10	3
105-mm HOWITZER M102, TOWED	11,500 14,500 (RAP)	0	10	3
155-mm HOWITZER M114A1, TOWED	14,600 14,600	0	4	1
155-mm HOWITZER M114A2, TOWED	14,600 19,400 (RAP)	0	4	1
155-mm HOWITZER M198, TOWED	24,000 30,000 (RAP)	0	4	TEMPERATURE DEPENDENT
155-mm HOWITZER M109, SP	14,600	0	4	1
155-mm HOWITZER M109A1/A2/A3, SP	18,100 23,500 (RAP)	0	4	1*
175-mm GUN M107, SP	32,800	0	1.5	0.5
203-mm HOWITZER M115, TOWED	16,800	0	1.5	0.5
203-mm HOWITZER M110, SP	16,800	0	1.5	0.5
203-mm HOWITZER M110A1, SP	20,600	0	1.5	0.5
203-mm HOWITZER M110A2, SP	22,900	0	1.5	0.5
MORTARS				
60-mm MORTAR	3,490 (HE) 1,472 (WP) 931 (ILLUM)	70 (HE) 33 (WP) 725 (ILLUM)	30	20
81-mm MORTAR	4,595 (HE) 4,850 (HE), track 4,737 (WP)	72 (HE) 70 (WP) 100 (ILLUM)	30 30	20 FOR 2 MINUTES, THEN 8 MINUTES
107-mm MORTAR	6,840 (HE) 5,650 (WP) 5,490 (ILLUM)	770 (HE) 920 (WP) 400 (ILLUM)	18	9 FOR 5 MINUTES, THEN 3 MINUTES

* CHG: ONE ROUND PER MINUTE FOR 60 MINUTES, THEN ONE ROUND EVERY 3 MINUTES THEREAFTER.

Table 9-1. Indirect-weapon systems capabilities.

WEAPON	MAXIMUM RANGE (METERS)	MINIMUM RANGE (METERS)	MAXIMUM RATE (ROUNDS PER MINUTE FIRST MINUTE)	SUSTAINED RATE (ROUNDS PER MINUTE)
NAVAL GUNFIRE				
5-inch/38	15,000	0	20	15
5-inch/54	22,500	0	30	20
16-inch/50	37,000	0	1	1

WEAPON	TRAVERSE LIMITS (mils)	SHELL/FUZE COMBINATIONS**
FIELD ARTILLERY		
105-mm HOWITZER	409R/400L	APICM, HE/PD, HE/DELAY, HE TRAINING INERT, HE/VT, HE/CP, RAP/PD, RAP/DELAY, WP/PD, WP/DELAY, WP/TRAINING INERT, SMOKE, ILLUM
105-mm HOWITZER M102, TOWED	6400	SAME AS ABOVE
155-mm HOWITZER M114A1, TOWED	448R/418L	CLGP, APICM, HE/PD, HE/DELAY, HE/TRAINING INERT, HE/VT, HE/CP, WP/PD, WP/DELAY, WP/TRAINING INERT, SMOKE, COLORED SMOKE, ILLUM
155-mm HOWITZER M114A2, TOWED	SAME AS ABOVE	CLGP, DPICM, RAAMS, ADAM, APICM, HE/PD, HE/DELAY, HE/TRAINING INERT, HE/VT, HE/CP, RAP/PD, RAP/DELAY, WP/PD, WP/DELAY, WP/TRAINING INERT, SMOKE, COLORED SMOKE, ILLUM
155-mm HOWITZER M198, TOWED	400R/400L 6400 SPEED	SAME AS ABOVE
155-mm HOWITZER M109, SP	6400	SAME AS ABOVE
155-mm HOWITZER M109A1/2/3, SP	6400	SAME AS ABOVE
175-mm GUN M107, SP	553R/533L	HE/PD, HE/DELAY, HE/TRAINING INERT, HE/VT
203-mm HOWITZER M115, TOWED	SAME AS ABOVE	APICM, HE/PD, HE/DELAY, HE/TRAINING INERT, HE/VT, HE/CP
203-mm HOWITZER M110, SP	SAME AS ABOVE	SAME AS ABOVE
203-mm HOWITZER M110A1, SP	SAME AS ABOVE	DPICM, APICM, HE/PD, HE/DELAY, HE/TRAINING INERT, HE/VT, RAP/PD, RAP/DELAY
203-mm HOWITZER M110A2, SP	SAME AS ABOVE	SAME AS ABOVE

** THESE REFLECT ONLY THOSE SHELL/FUZE COMBINATIONS THE OBSERVER MAY REQUEST— NOT ALL THOSE AVAILABLE.

Table 9-1. Indirect-weapon systems capabilities (continued).

WEAPON	TRAVERSE LIMITS (mils)	SHELL/FUZE COMBINATIONS**
MORTARS		
60-mm MORTAR	250R/250L	HE/PD, HE/DELAY, HE/VT, WP/PD, WP/DELAY, WP/TRAINING INERT, ILLUM
81-mm MORTAR	95R/95L 6400 TRACK	HE/PD, HE/DELAY, HE/VT, WP/PD, WP/DELAY, ILLUM
107-mm MORTAR	125R/125L 6400 TRACK	HE/PD, HE/DELAY, HE/TRAINING INERT, HE/VT, WP/PD, WP/DELAY, ILLUM
120-mm MORTAR		
NAVAL GUNFIRE		
5-inch/38	6400***	HE/PD, HE/TRAINING INERT, HE/VT, HE/CP, AP/DELAY, WP/PD, WP/TRAINING INERT, ILLUM
5-inch/54	6400***	SAME AS ABOVE
167-inch/50	6400***	HE/PD, HE/TRAINING INERT, HE/VT, HE/CP, AP/DELAY

  * CHG: ONE ROUND PER MINUTE FOR 60 MINUTES, THEN ONE ROUND EVERY 3 MINUTES THEREAFTER.

  ** THESE REFLECT ONLY THOSE SHELL/FUZE COMBINATIONS THE OBSERVER MAY REQUEST— NOT ALL THOSE AVAILABLE.

  *** WITH INCREASED MINIMUM RANGES WHEN FIRING OVER SHIP'S STRUCTURES.

Table 9-1. Indirect-weapon systems capabilities (continued).

b. **Insertion/Extraction Techniques.** Practical application of insertion/extraction techniques enables snipers to accomplish its mission and to exfiltrate with confidence. Leaders should tailor these techniques to unit assets; however, a working knowledge of all techniques listed in Chapter 7 is an invaluable tool to the team.

c. **Tracking/Counterattacking.** Footprints found by enemy trackers may indicate that snipers are in the area. A knowledge of countertracking techniques is a valuable tool to snipers not only to remain undetected but also to collect battlefield information. (See Chapter 8.)

d. **Survival Skills.** Survival training, incorporated with evasion and escape training, will better prepare the sniper in contingency planning during exfiltration and, possibly, infiltration. Judging enemy reaction is an impossible task therefore, the sniper may be forced to live off the land until linkup can be established with friendly forces.

e. **First Aid.** Adequate first-aid training can mean the difference between life and death until proper medical attention can be given.

f. **Communications Reporting Procedures.** A lack of timely, detailed reporting of battlefield information can hinder the overall success of maneuvering units. Properly formatted information (Chapter 6), precoordinated with communications personnel, ensures timely and accurate intelligence gathering. Snipers must train to use information reporting formats and procedures.

9-3. TRAINING NOTES

Snipers should be trained IAW DA Pamphlet 350-38. Training includes knowledge of equipment, ammunition, range and terrain requirements, and techniques of training and sustaining the skills of the sniper team.

a. **Equipment.** During all FIXs, each sniper should be equipped as indicated in Chapter 2. Team equipment should be available as needed.

b. **Known Distance Range Requirements.** A standard known-distance range, graduated in 100-meter increments from 100 to 1,000 meters, is required for zeroing and zero confirmation exercises. The target detection range facilities and procedures should permit observation and range determination to 800 meters.

c. **Field Firing Range Requirements.** The ideal field firing range should be on terrain that has been left in its natural state. The range should be a minimum of 800 meters in depth with provisions along the firing line for several sniper positions within each lane to provide a slightly different perspective of the target area (Table 9-2). Where time prevents construction of a separate range, it may be necessary to superimpose this facility over an existing field firing range.

(1) Iron maidens can be made out of 3/4-inch steel plate with a supporting frame. They should be cut out in the form of silhouettes 20 inches wide and 40 inches high. By painting these targets white, the sniper can easily detect where the bullet impacts on the target.

(2) Placing targets inside of window openings gives the sniper experience engaging targets that can be found in an urban environment. This is done by cutting a 15-inch by 15-inch hole in the center of a 36-inch by 48-inch plywood board. Then an E-type silhouette is emplaced on a hit-kill mechanism 2 to 4 meters behind the plywood.

(3) Targets placed inside a bunker-type position allows the sniper to gain experience firing into darkened openings. This position can be built with logs and sandbags with an E-type silhouette on a hit-kill mechanism placed inside.

METERS	TYPE TARGET
200	E-TYPE SILHOUETTE, HIT-KILL MECHANISM.
300	IRON MAIDEN SILHOUETTE; E-TYPE SILHOUETTE, HIT-KILL MECHANISM; MOVING TARGET MECHANISM.
325	E-TYPE SILHOUETTE, HIT-KILL MECHANISM.
375	E-TYPE SILHOUETTE, HIT-KILL MECHANISM, EMPLACED INSIDE A WINDOW.
400	E-TYPE SILHOUETTE, HIT-KILL MECHANISM, EMPLACED INSIDE A BUNKER.
500	IRON MAIDEN SILHOUETTE; MOVING TARGET MECHANISM, TRACKED VEHICLE WITH A HIT-KILL MECHANISM IN THE COMMANDER'S CUPOLA.
600 to 1,000	IRON MAIDEN SILHOUETTES.

Table 9-2. Field firing range requirements.

(4) Moving targets can be used at distances between 300 and 500 meters to give the sniper practical experience and to develop skill in engaging a moving target. Two targets, one moving laterally and one moving at an oblique, present a challenge to the sniper.

(5) Targets should be arranged to provide varying degrees of concealment to show enemy personnel or situations in logical locations (Figure 9-9, page 9-24). The grouping of two or more targets to indicate a crew-served weapon situation or a small unit is acceptable. Such arrangements, provided the targets can be marked, may require selective engagement by the sniper. The automatic target devices provide for efficient range operation and scoring.

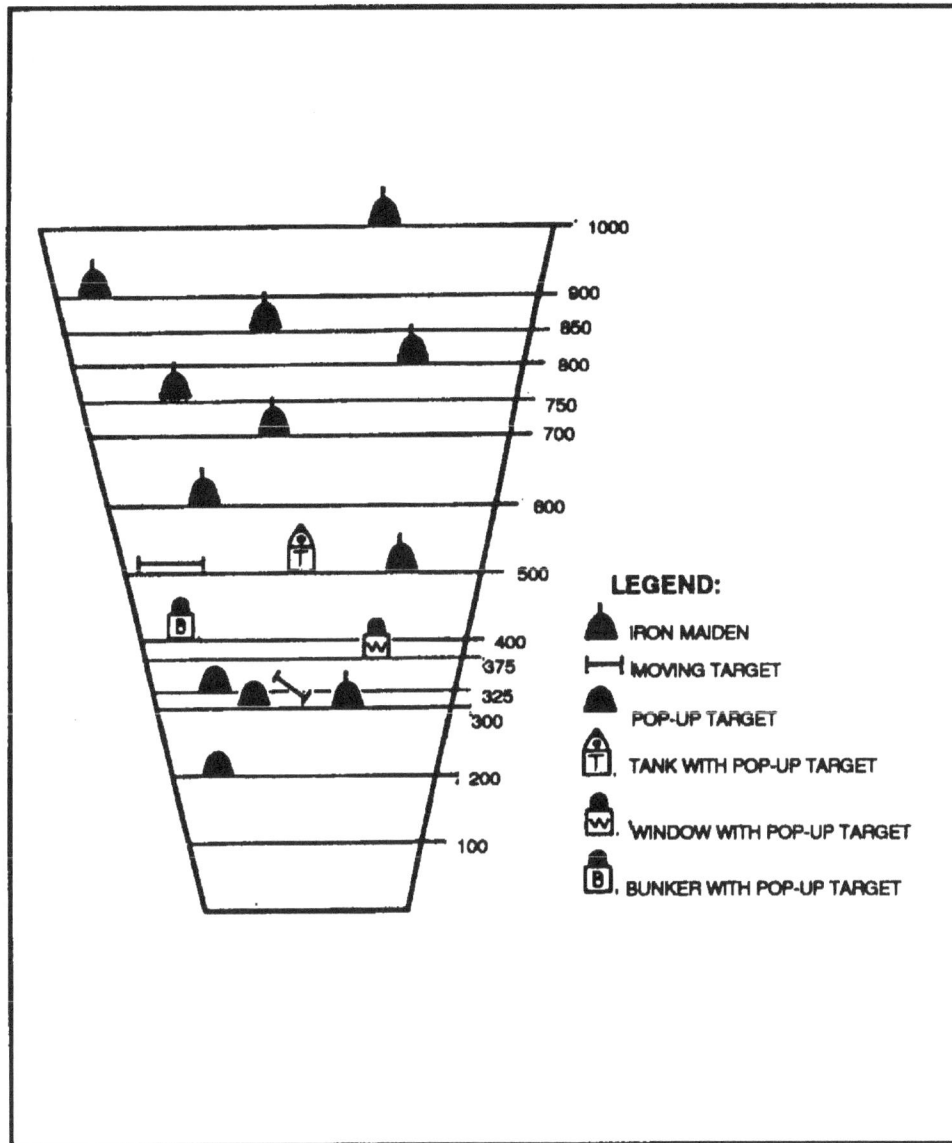

Figure 9-9. Lane layout.

9-4. EXAMPLE 5-DAY SNIPER SUSTAINMENT TRAINING PROGRAM

An example of a 5-day sniper sustainment training program is as follows:

DAY 1

TASK 1: Select sniper team routes and positions.

CONDITIONS: Given a review of selection of routes and positions, a situational sniper mission with a target area location that requires a minimum movement of 3,000 meters, a military map, a protractor, a felt-tip pen, an 8-inch-square clear plastic overlay, and one sheet of letter-size paper.

STANDARDS: Select and plot a primary and alternate route, objective rally point, and tentative final firing position that provides the best cover and concealment.

    1. Prepare overlay with two grid reference marks; primary and alternate routes with arrows indicating direction of travel; minimum of three checkpoints, numbered in order; ORP; and a tentative final firing position.

    2. Prepare a written log of movement. The sniper data book will contain the from-to grid coordinates, magnetic azimuths, distance, checkpoint number, objective rally point, and tentative final firing position.

    3. Prepare overlay and written log of movement within 30 minutes.

TASK 2: Move while using individual sniper movement techniques.

CONDITIONS: Given a review of sniper movement techniques, a sniper weapon, a ghillie suit, and a flat, open area that allows trainers to observe movement techniques.

STANDARDS: Move correctly while using the designated movement technique.

    1. Sniper low crawl.

    2. Medium crawl.

    3. High crawl.

    4. Hands-and-knee crawl.

    5. Walking.

NOTE: Trainers designate movement techniques and critique snipers on their movement.

TASK 3: React to enemy contact while moving as a member of a sniper team.

CONDITIONS: Given a review of sniper team movement techniques and reactions to enemy contact, sniper team's basic equipment and weapons, and an area of varying terrain with at least one danger area.

STANDARDS: React correctly to designated situations or danger areas.

    1. Visual contact.

    2. Ambush.

    3. Indirect fire.

    4. Air attack.

    5. Danger area (linear and open area).

NOTE: Trainers designate situations and critique sniper teams on movement.

TASK 4: Describe target detection, selection, and observation techniques.

CONDITIONS: Given a review of target detection, selection, and observation techniques.

STANDARDS: Describe, orally or in writing, techniques used to observe, detect, and select targets.

TASK 5: Identify enemy uniforms, equipment, and vehicles.

CONDITIONS: Given a review of pictures or slides of enemy uniforms, equipment, and vehicles.

STANDARDS: Identify 7 of 10 enemy uniforms or rank insignia, 7 of 10 pieces of enemy equipment, and 7 of 10 enemy vehicles.

TASK 6: Describe range estimation techniques.

CONDITIONS: Given a review of range estimation techniques used by snipers.

STANDARDS: Describe, orally or in writing, range estimation techniques used by the sniper.

    1. Eye methods.

    2. Use of binoculars.

    3. Use of M3A scope/ M49 observation telescope.

TASK 7: Prepare a sniper range card.

CONDITIONS: Given a review of sniper range cards, a suitable target area, basic sniper equipment, and a sniper range card.

STANDARDS: Prepare a sniper range card complete with—

1. Grid coordinates of position.

2. Target reference point(s) (azimuth, distance, and description).

3. Left/ right limits with azimuths.

4. Ranges throughout area.

5. Major terrain features.

6. Method of obtaining range/ name.

7. Weather data.

TASK 8: Prepare a military sketch.

CONDITIONS: Given a review of sniper military sketching, a suitable area or object to sketch, and a blank military sketch sheet.

STANDARDS: Prepare a sketch complete with—

1. Grid coordinates of position.

2. Magnetic azimuth through center of sketch.

3. Sketch name and number.

4. Scale of sketch.

5. Remarks section.

6. Name/ rank.

7. Date/ time.

8. Weather data.

TASK 9: Maintain a sniper data book.

CONDITIONS: Given a review of the sniper data book and 20 blank sheets stapled together as a booklet.

STANDARDS: Maintain a sniper data book with a chronological listing of events that take place during the next three days and containing the following:

1. Grid coordinates of position.

2. Observer's name.

3. Date/ time/ visibility.

4. Sheet number/ number of total sheets.

5. Series number/ time and grid coordinate of each event.

6. Event.

7. Action taken.

NOTE: Trainers collect the sniper data books in three days.

DAY 2

TASK 1: Describe the fundamentals of sniper marksmanship.

CONDITIONS: Given a review of sniper marksmanship fundamentals.

STANDARDS: Describe, orally or in writing, the fundamentals of sniper marksmanship.

1. Position.

2. Breath control.

3. Aiming.

4. Trigger control.

TASK 2: Describe the effects of weather on ballistics.

CONDITIONS: Given a review of the effects of weather on ballistics.

STANDARDS: Describe, orally or in writing, the effects of weather on ballistics.

TASK 3: Describe the sniper team method of engaging targets.

CONDITIONS: Given a review of the sniper team method of engaging targets.

STANDARDS: Describe, orally or in writing, the sniper team method of engaging targets.

TASK 4: Describe methods used to engage moving targets.

CONDITIONS: Given a review of methods used to engage moving targets.

STANDARDS: Describe, orally or in writing, methods used to engage moving targets.

TASK 5: Describe methods used to engage targets at various ranges without adjusting the scope's elevation.

CONDITIONS: Given a review of methods used to engage targets at various ranges without adjusting the scope's elevation.

STANDARDS: Describe, orally or in writing, the methods used to engage targets at various ranges without adjusting the scope's elevation.

TASK 6: Zero rifle scope.

CONDITIONS: Given a sniper weapon, an M49 observation telescope, a suitable firing range, and 7 rounds of 7.62-mm special ball (Ml 18) ammunition.

STANDARDS: Zero rifle scope within 7 rounds.

DAY 3

TASK 1: Zero iron sights.

CONDITIONS: Given a sniper weapon, a suitable firing range, and 12 rounds of 7.62-mm special ball ammunition.

STANDARDS: Zero iron sights on a sniper weapon within 12 rounds.

TASK 2: Engage moving targets.

CONDITIONS: Given a sniper weapon, an M49 observation telescope, a suitable firing range, and 10 rounds of 7.62-mm special ball (M118) ammunition.

STANDARDS: Engage 10 moving targets, from 300 to 500 meters, achieving a minimum of 7 hits.

TASK 3: Estimate range.

CONDITIONS: Given a sniper weapon system (M24), M19 binoculars, and 10 targets out to 800 meters.

STANDARDS: Correctly estimate range to 7 of the 10 targets using eye estimation (± 15 percent), binoculars (± 10 percent), or the M24 sniper weapon (± 5 percent).

TASK 4: Detect targets.

CONDITIONS: Given a suitable area with 10 military objects, binoculars, M49 observation telescope, and a scorecard.

STANDARDS: Detect, plot, and describe 7 of 10 military objects within 40 minutes.

TASK 5: Participate in a concealment exercise.

CONDITIONS: Given a sniper weapon, ghillie suit, three 7.62-mm blank rounds of ammunition, an area to conceal a sniper position, and 10 minutes to prepare.

STANDARDS: Without being detected, occupy a position, identify, and fire three blank rounds at a target (located 100 to 200 meters away) who is equipped with binoculars and an M49 observation telescope. Must score 7 of 10 points (Figure 9-10).

IF THE SNIPER—	POINTS		
	GIVEN	DEDUCTED	TOTAL
WAS DETECTED WITHOUT THE AID OF OPTICS (FIRST 2 MINUTES)	2	0	2
WAS DETECTED WITH THE AID OF OPTICS (18 MINUTES)	1	0	3
WAS DETECTED WHEN ASSISTANT TRAINER IS WITHIN 10 FEET OF SNIPER	1	0	4
PROPERLY IDENTIFIED THE NUMBER WITHIN 30 SECONDS	1	0	5
FAILED TO PROPERLY IDENTIFY THE NUMBER	0	3	2
FIRED FIRST SHOT, NOT DETECTED	4	0	6
FIRED SECOND SHOT, NOT DETECTED	1	0	7
MAINTAINED STABLE FIRING POSITION (SUPPORT)	2	0	9
PROPERLY ADJUSTED WEAPON'S SCOPE FOR RANGE AND WINDAGE	1	0	10

NOTES: 1. IF THE SNIPER IS CAUGHT TRYING TO IDENTIFY THE NUMBER SCORE 4 POINTS.

2. IF MUZZLE BLAST/FLASH IS DETECTED, DEDUCT 1 POINT FROM TOTAL SCORE.

3. FAILING TO COMPLY WITH TRAINING STANDARDS AND OBJECTIVES (SUCH AS UNNECESSARY MOVEMENT, PREMATURE FIRE, OUTSIDE OF PRESCRIBED BOUNDARIES) WILL RESULT IN TERMINATION OF THE EXERCISE AND A SCORE OF ZERO.

DAY 4

TASK 1: Quality on Qualification Table No. 1.

CONDITIONS: Given a sniper weapon, M49 observation telescope, a suitable firing range, Qualification Table No. 1 scorecard, and 40 rounds of 7.62-mm special ball (Ml 18) ammunition.

STANDARDS: Engage targets from 200 to 700 meters, achieving a minimum of 140 points.

TASK 2: Engage targets in MOPP.

CONDITIONS: During daylight, given a sniper weapon, suitable firing range, MOPP suit, complete M25-series protective mask, M49 observation telescope, and 30 rounds of 7.62-mm special ball (M118) ammunition.

STANDARDS: While in MOPP, engage targets at 300 to 800 meters, achieving a minimum of 105 points.

TASK 3: Participate in a concealed movement exercise.

CONDITIONS: Given a sniper weapon, ghillie suit, two 7.62-mm blank rounds of ammunition, and a suitable area 1,000 meters long that is observable.

STANDARDS: Within 4 hours, move 600 to 800 meters; without being detected, occupy a position, identify, and fire two blank rounds at an enemy target who is equipped with binoculars and an M49 observation telescope. Must score 7 of 10 points (Figure 9-11).

IF THE SNIPER—	POINTS		
	GIVEN	DEDUCTED	TOTAL
FAILED TO CROSS THE FFL	0	0	0
CROSSED THE FFL	6	0	6
FIRED FIRST SHOT, NOT DETECTED	2	0	8
WAS NOT DETECTED WHEN ASSISTANT TRAINER IS WITHIN 10 FEET OF SNIPER	2	0	10
PROPERLY IDENTIFIED THE 1ST NUMBER	2	0	12
WAS NOT DETECTED WHEN ASSISTANT TRAINER IS WITHIN 5 FEET OF SNIPER	2	0	14
FIRED SECOND SHOT, NOT DETECTED	2	0	16
PROPERLY IDENTIFIED THE 2D NUMBER	2	0	18
MAINTAINED GOOD CAMOUFLAGE	1	0	19
MAINTAINED STABLE FIRING POSITION (SUPPORT)	1	0	20

NOTES: 1. IF MUZZLE BLAST/FLASH IS DETECTED, DEDUCT 1 POINT FROM THE TOTAL SCORE.

2. FAILING TO COMPLY WITH TRAINING STANDARDS AND OBJECTIVES (SUCH AS UNNECESSARY MOVEMENT, PREMATURE FIRE, OUTSIDE OF PRESCRIBED BOUNDARIES) WILL RESULT IN TERMINATION OF THE EXERCISE AND A SCORE OF ZERO.

Figure 9-11. Scoring for concealed movement exercise.

DAY 5

TASK 1: Qualify on Qualification Table No. 2.

CONDITIONS: Given a sniper weapon, M49 observation telescope, a suitable firing range, Qualification Table No. 2 scorecard, and 40 rounds of 7.62-mm special ball (Ml18) ammunition.

STANDARDS: Engage targets at 300 to 900 meters, achieving a minimum of 140 points.

TASK 2: Call for fire.

CONDITIONS: Given a review of call-for-fire procedures, two AN/ PRC-77 radios, and a fire mission.

STANDARDS: Transmit the fire mission using proper radio procedures and the elements of the call-for-fire mission in sequence:

1. Observer identification.

2. Warning order.

3. Target location.

4. Target description.

5. Method of engagement (optional).

6. Method of fire and control (optional).

TASK 3: Locate target by grid coordinates.

CONDITIONS: Given a review of locating targets using the grid-coordinate method, a map of the target area, binoculars, compass, and a target.

STANDARDS: Determine and announce the six-digit coordinates of the target (within a 250-meter tolerance) within 30 seconds.

TASK 4: Locate a target by polar plot.

CONDITIONS: Given a review of target locating using the polar-plot method, a map of the target area, binoculars, a compass, and a target.

STANDARDS: Locate the target within 250 meters of the actual location. Announce the target location within 30 seconds after identification. Express direction to the nearest 10 roils and within 100 mils of actual direction. Express distance to the nearest 100 meters.

TASK 5: Locate target by shift from a known point.

CONDITIONS: Given a review of locating targets using the shift from a known-point method, a map of the target area, binoculars, a compass, a known point, and a target.

STANDARDS: Locate the target within 250 meters of the actual location and announce the target location within 30 seconds after identification. Express direction to the nearest 10 roils and within 100 roils of the actual direction. Express right or left corrections to the nearest 10 meters and range corrections to the nearest 100 meters.

TASK 6: Participate in a land navigation exercise during daylight.

CONDITIONS: Given a navigation course with at least four legs no less than 800 meters apart.

STANDARDS: Navigate the course without being detected by the observer-instructor. Preparing sketches, range cards, and or logs from the sniper data book can also be incorporated into the exercise.

NIGHT 5

TASK: Participate in a land navigation exercise during nightfall.

CONDITIONS: Given a navigation course (FM 21-26) with at least three legs no less than 500 meters apart. Observer-instructors can be placed on the course to detect any violations of noise and light discipline and deduct points from the sniper's score for violations.

STANDARDS: Navigate the course without being detected.

9-5. EMERGENCY DEPLOYMENT READINESS EXERCISE

Trainers use T&EOs from ARTEP 7-92-MTP: Move Tactically (7-5-1825); Select/ Engage Targets (7-5-1869); Select/ Occupy Firing Position (7-5-1871); Estimate Range (7-5-1872); and Debrief (7-5-1809) for additional sustainment training. An example of a battalion EDRE follows:

TIME	ACTION
0400	Battalion alerts sniper teams.
	1. CQ relays uniform and packing list.
	2. Sniper teams have two hours to report to battalion.
	3. Sniper team leaders report to SEO when all of the team is accounted for.
	4. Sniper team receives FRAGO from the SEO.
0600	Snipers depart battalion area by air, truck, or road march.
0800	Sniper teams arrive at range.
	1. Sniper teams receive range/ safety briefing.
	2. Snipers receive issued ammunition.
	3. Snipers zero weapons.
	4. Sniper teams field/ record fire on a range with targets positioned from 200 to 900 meters.
1100	Sniper teams depart range; move to concealed movement site by truck, road march, or tactical movement by teams.

TIME	ACTION
1200	Sniper teams arrive at conceded movement site.

1. Sniper teams receive briefing.

2. Site should be 800 to 1,000 meters long positioned with a observer- instructor as a target at one end with field table, M19 binoculars, M49 observation telescope, 8-inch by 8-inch international orange panels with white 5-inch number (1 to 9) painted on them, and two AN/ PRC-77 radios for observer and assistant trainer.

3. Sniper will have four hours to move into his FFP, 50 to 200 meters from observer-instructor, and fire his first shot.

4. Sniper will have 30 seconds in which to identify number.

5. Sniper will fire second shot.

NOTE: All information is to be recorded in the sniper data book.

6. The entire exercise will be conducted without the sniper being detected by the observer-instructor.

1600   Sniper teams depart for day/ night land navigation exercise.

1. Sniper teams start the exercise from a concealed movement site.

2. Sniper teams will be required to move to three different points. At each point they will perform one of the following

- Draw a militaty sketch.
- Draw a range card.
- Do a target detection exercise.
- Collect information/ data.

3. All movement will be performed without being detected.

2000   Night navigation exercise.

1. Sniper teams start the exercise from the command post.

TIME	ACTION
	2. They will move undetected to three different points.
	3. They will perform a detection exercise with the use of NODS.
	4. They will record all information in the sniper data book.
	5. After collecting necessary data, they will move to an extraction point and construct a sniper hide position. They will prepare for target reduction.
0500-0600	Target reduction.
	1. Upon target reduction time, the sniper team will prepare for extraction.
	2. At extraction time, they will return to the battalion area.
	3. The SEO will debrief the sniper team.
	4. The SEO will conduct an after-action review.

NOTE: A written test could also be given as part of the EDRE.

9-6. RECORD FIRE TABLES

In accordance with DA Pamphlet 350-38, sniper qualification should occur quarterly. Sniper qualification involves the firing of two field fire tables. Qualification Table No. 1 grades target engagements primarily between 200 and 700 meters. Scores are recorded on DA Form 7329-R, Qualification Table No. 1 Scorecard (Figure 9-12, page 9-38). Qualification Table No. 2 grades on the longer ranges between 300 to 900 meters. Scores are recorded on DA Form 7330-R, Qualification Table No. 2 Scorecard (Figure 9-13, page 9-39). Although the sniper weapon system has an 800-meter maximum effective range, it can effectively hit targets at 1,000 meters. This is a challenge to the sniper and, with successful engagement, is a confidence builder in his ability. To qualify on firing tables No. 1 and No. 2, the sniper must adhere to the following standards:

NOTE: Completion of the DA Forms 7329-R and 7330-R is self-explanatory. Blank copies of these forms are located at the back of this manual for local reproduction.

- Achieve a 70 percent standard of 140 points out of a possible 200 points.

- Fire a first-round hit to equal 10 points.
- Fire a second round if the first round misses the target.
- Receive 5 points if the second round hits the target.
- Receive a score of O if the second round misses the target.
- Complete firing within 30 minutes. Total all first-round hits and multiply by 10; total second-round hits and multiply by 5.
- Add first-round and second-round hits for a total firing table score.
- Meet the 70 percent standard (140 points). Trainer checks satisfactory or unsatisfactory.

NOTE: Trainer and sniper sign the scorecard.

Figure 9-12. Example of completed DA Form 7329-R, Qualification Table No. 1 Scorecard.

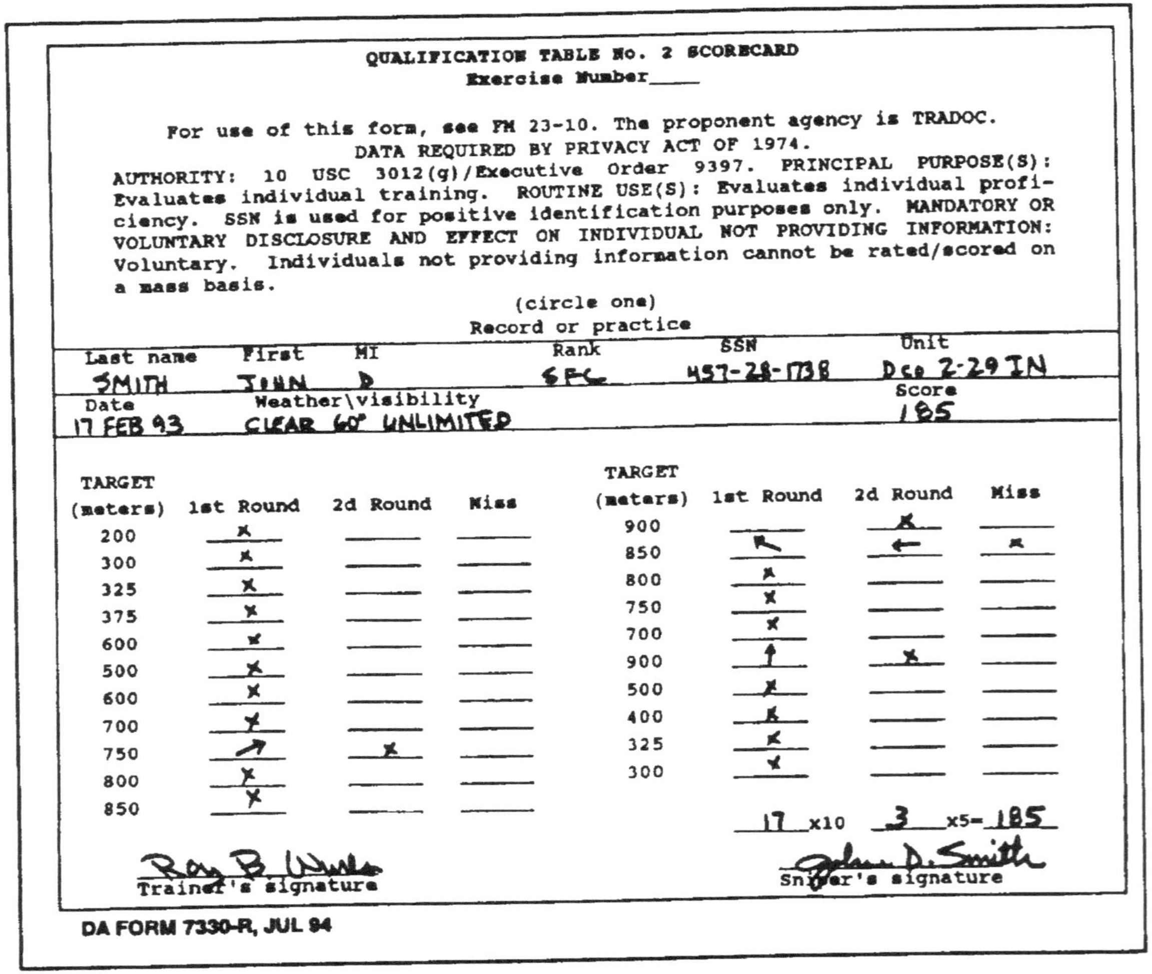

QUALIFICATION TABLE NO. 2 SCORECARD
Exercise Number _____

For use of this form, see FM 23-10. The proponent agency is TRADOC.
DATA REQUIRED BY PRIVACY ACT OF 1974.

AUTHORITY: 10 USC 3012(g)/Executive Order 9397. PRINCIPAL PURPOSE(S):
Evaluates individual training. ROUTINE USE(S): Evaluates individual profi-
ciency. SSN is used for positive identification purposes only. MANDATORY OR
VOLUNTARY DISCLOSURE AND EFFECT ON INDIVIDUAL NOT PROVIDING INFORMATION:
Voluntary. Individuals not providing information cannot be rated/scored on
a mass basis.

(circle one)
Record or practice

Last name	First	MI	Rank	SSN	Unit
SMITH	JOHN	D	SFC	457-24-1738	Dco 2-29 IN

Date	Weather\Visibility	Score
17 FEB 93	CLEAR 60" UNLIMITED	185

TARGET (meters)	1st Round	2d Round	Miss
200	X		
300	X		
325	X		
375	X		
600	X		
500	X		
600	X		
700	↗		
750	X	X	
800	X		
850			

TARGET (meters)	1st Round	2d Round	Miss
900	X	X	
850		↓	X
800	X		
750	X		
700	↑		
900	X	X	
500	X		
400	X		
325	X		
300	X		

17 x10   3   x5 = 185

Trainer's signature

Shooter's signature

DA FORM 7330-R, JUL 94

Figure 9-13. Example of completed DA Form 7330-R,
Qualification Table No. 2 Scorecard.

9-7. M24 SNIPER MILES TRAINING

MILES training is an invaluable tool to simulate realistic combat training. Other than actual combat, the sniper's best means of displaying effectiveness as a force multiplier is through the use of the M24 sniper weapon system (MILES).

a. **Characteristics of the MILES Transmitter.** The M24 sniper weapon system MILES transmitter is a modified M16 transmitter. A special mounting bracket attaches the laser transmitter to the right side of the barrel (looking from the butt end) of the M24 and places it parallel with the line of bore. The laser beam output has been amplified and tightened to provide precision fire capability out to 1,000 meters. (For component information and instructions on mounting, zeroing, and operation, see TM 9-1265-211-10.)

b. **Training Value.** Using the M24 MILES, the trainer can enhance sustainment training in target engagement.

(1) *Selection of firing positions.* Due to transmitter modifications, the sniper must attain a firing position that affords clear fields of fire. Any obstruction (vegetation, terrain) can prevent a one-shot kill by deflecting or blocking the path of the laser beam. By attaining this type of position, the sniper will improve his observation and firing capabilities.

(2) *Target detection/selection.* Using MILES against multiple/ cluster targets requires the sniper to select the target that will have the greatest effect on the enemy. The trainer provides instant feedback on the sniper's performance. Situations may be created such as bunkers, hostage situations, and MOUT firing. The hit-miss indicating aspects of MILES are invaluable in this type of training.

(3) *Range estimation.* The sniper must be highly skilled in range estimation (Chapter 3) to properly use the M24 sniper weapon system. The trainer's evaluation of this ability is as simple as the sniper pulling the trigger. When the range to the target is properly computed and elevation dialed on the M3A, one shot, either hit or miss, indicates a strength or weakness in the sniper's range estimation ability (if the fundamentals of marksmanship were properly applied).

(4) *Markmanship.* A target hit (kill) with MILES is the same as one with live ammunition. Applying marksmanship fundamentals results in a first-round kill; the training value is self-evident.

c. **MILES Training Limitations.** The concept of MILES is to provide realistic training however, MILES is limited in its capabilities as applied to the sniper's mission of long-range precision fire.

(1) *Lack of external ballistics training.* A laser is a concentrated beam of light emitted by the MILES transmitter. It travels from the sniper's

weapon undisturbed by outside forces such as temperature, humidity, and wind. Lack of these effects may lull the sniper into a false sense of confidence. The trainer should constantly reinforce the importance of these factors. The sniper should make a mental note of changes that should be applied to compensate for these effects.

(2) *Engagement of moving targets.* The engagement of moving targets (Chapter 3) requires the sniper to establish a target lead to compensate for flight time of his bullet. Traveling in excess of 186,000 miles per second (speed of light), the MILES laser nullifies the requirement for target lead. Again, the sniper may be lulled into a false sense of confidence. The trainer should enforce the principles of moving target engagement by having the sniper note appropriate target lead for the given situation.

APPENDIX A

PRIMARY SNIPER WEAPONS OF THE WORLD

Several countries have developed sniper weapon systems comparable to the United States systems. These weapon systems are sold to or copied by countries throughout the world. Within the everchanging world of politics, it is impossible to predict how the future enemy may be armed. The designs and capabilities of these weapon systems are sirnilar. However, the amount of training and experience separates the sniper the marksman. This appendix describes the characteristics and capabilities of prevalent sniper weapon systems.

A-1. AUSTRIA

The Austrian Scharfschutzengewehr 69 (SSG-69) is the current sniper weapon of the Austrian Army and several foreign military forces. It is available in either 7.62-mm x 51 or the .243 Winchester calibers. The SSG-69 is a manually bolt-operated, 5-round rotary or 10-round box, magazine-fed, single-shot repeating rifle. Recognizable features are synthetic stock hammer-forged, heavy barrel with a taper; two-stage trigger, adjustable for length and weight of pull; and a machined, longitudinal rib on top of the receiver that accepts all types of mounts. The sighting system consists of the Kahles ZF69 6-power telescope iron sights are permanently affixed to the rifle for emergency use. The telescope comes equipped with an internal bullet-drop compensator graduated to 800 meters, and a reticle that consists of an inverted V with broken cross hairs. The weapon, magazine, and telescope together weigh 10.14 pounds. This weapon has a barrel length of 25.59 inches and a total length of 44.88 inches with a muzzle velocity of 2,819 feet per second. It has an accuracy of 15.75 inches at 800 meters using RWS Match rounds.

A-2. BELGIUM

The Model 30-11 sniping FN rifle is the current sniper rifle of the Belgian and other armies. This weapon is a 7.62-mm x 51, 5-round internal or 10-round detachable box, magazine-fed, manually bolt-operated rifle with a Mauser-action heavy barrel and, through the use of butt-spacer plates, an adjustable stock. Its sighting system is the FN 4-power, 28-mm telescope and aperture sights with 1/6 MOA adjustment capability. The rifle weighs 10.69 pounds and, with its 19.76-inch barrel, is a total of 43.97 inches long. The Model 30-11 has a muzzle velocity of 2,819 fps. Accessories include the biped of the MAG machine gun, butt-spacer plates, sling, and carrying case.

A-3. THE FORMER CZECHOSLOVAKIA

The current sniper weapon system is the VZ54 sniper rifle. It is a manually bolt-operated, 10-round box, magazine-fed 7.62-mm x 54 rimmed weapon and built upon bolt-action with a free-floating barrel. This weapon is similar to the M1891/30 sniping rifle (former Russian weapon)-only shorter and lighter. The rifle is 45.19 inches long and weighs 9.02 pounds with the telescope. It has a muzzle velocity of 2,659 fps with a maximum effective range of 1,000 meters.

A-4. FINLAND

Finnish weapon technology introduces a 7.62-mm x 51 sniper rifle that is equipped with an integral barrel/silencer assembly. It is a bolt-action, 5-round box, magazine-fed weapon with a nonreflective plastic stock and a standard adjustable biped. Through the use of adaptors, any telescopic or electro-optical sight may be mounted. The weapon is not equipped with metallic sights. The 7.62-mm Vaime SSR-1 (silenced sniper rifle) weighs 9.03 pounds and is 46.45 inches long.

A-5. FRANCE

French sniper weapons consist of the FR-F1 and FR-F2.

   a. **FR-F1.** The FR-F1 sniping rifle, known as the Tireur d'Elite, is a manually bolt-operated, 10-round detachable box, magazine-fed, 7.62-mm x 51 or 7.5-mm x 54 weapon. The length of the stock may be adjusted with the butt-spacer plates. This weapon's sighting system consists of the Model 53 bis 4-power telescopic sight and integral metallic sights with luminous spots for night firing. It weighs 11.9 pounds, has a barrel length of 21.7 inches, and a total length of 44.8 inches. This weapon has a muzzle velocity of 2,794 fps and a maximum effective range of 800 meters.

Standard equipment features a permanently affixed biped whose legs may be folded forward into recesses in the fore-end of the weapon.

b. **FR-F2.** The FR-F2 sniping rifle is an updated version of the F1. Dimensions and operating characteristics remain unchanged; however, functional improvements have been made. A heavy-duty biped has been mounted more toward the butt-end of the rifle, adding ease of adjustment for the firer. Also, the major change is the addition of a thick, plastic thermal sleeve around and along the length of the barrel. This addition eliminates or reduces barrel mirage and heat signature. It is also chambered for 7.62-mm x 51 NATO ammunition.

A-6. GERMANY

The FRG has three weapons designed mainly for sniping the Model SP66 Mauser, WA 2000 Walther, and Heckler and Koch PSG-1.

a. **Model SP66 Mauser.** The SP66 is not only used by the Germans but also by about 12 other countries. This weapon is a heavy-barreled, manually bolt-operated weapon built upon a Mauser short-action. Its 26.8-inch barrel, completely adjustable thumbhole-type stock, and optical telescopic sight provide a good-quality target rifle. The weapon has a 3-round internal magazine fitted for 7.62-mm x 51 ammunition and a Zeiss-Diavari ZA 1.5-6-variable power x 42-mm zoom telescopic sight. The muzzle of the weapon is equipped with a flash suppressor and muzzle brake.

b. **WA 2000 Walther.** The WA 2000 is built specifically for sniping. The entire weapon is built around the 25.6-inch barrel; it is 35.6 inches long. This uniquely designed weapon is chambered for .300 Winchester Magnum, but it can be equipped to accommodate 7.62-mm x 51 NATO or 7.5-mm x 55 Swiss calibers. It is a gas-operated, 6-round box, magazine-fed weapon, and it weighs 18.3 pounds. The weapon's trigger is a single- or two-staged type, and its optics consist of a 2.5-10-variable power x 56-mm telescope. It has range settings of 100 to 600 meters and can be dismounted and mounted without loss of zero.

c. **Heckler and Koch PSG-1.** The PSG-1 is a gas-operated, 5- or 20-round, magazine-fed, semiautomatic weapon and is 47.5-inches long with a 25.6-inch barrel and has a fully adjustable, pistol-grip-style stock. Optics consist of a 6-power x 42-mm telescopic sight with six settings for range from 100 to 600 meters. The 7.62-mm x 51 PSG-1 weighs 20.7 pounds with tripod and when fully loaded. The muzzle velocity is 2,558 to 2,624 fps.

A-7. ISRAEL

The Israelis copied the basic operational characteristics and configuration of the 7.62-mm Galil assault rifle and developed a weapon to meet the demands of sniping. The 7.62-mm x 51 Galil sniping rifle is a semiautomatic, gas-operated, 20-round bolt magazine-fed weapon. Like most service rifles modified for sniper use, the weapon is equipped with a heavier barrel fitted with a flash suppressor it can be equipped with a silencer that fires subsonic ammunition. The weapon features a pistol-grip-style stock, a fully adjustable cheekpiece, a rubber recoil pad, a two-stage trigger, and an adjustable biped mounted to the rear of the fore-end of the rifle. Its sighting system consists of a side-mounted 6-power x 40-mm telescope and fixed metallic sights. The weapon is 43.89-inches long with a 20-inch barrel without a flash suppressor and weighs 17.64 pounds with a biped, sling, telescope, and loaded magazine. When firing FN Match ammunition, the weapon has a muzzle velocity of 2,672 fps; when firing M118 special ball ammunition, it has a muzzle velocity of 2,557 fps.

A-8. ITALY

The Italian sniper rifle is the Berretta rifle. This rifle is a manually bolt-operated, 5-round box, magazine-fed weapon, and fires the 7.62-mm x 51 NATO rounds. Its 45.9-inch length consists of a 23-inch heavy, free-floated barrel, a wooden thumbhole-type stock with a rubber recoil pad, and an adjustable cheekpiece. Target-quality, metallic sights consist of a hooded front sight and a fully adjustable, V-notch rearsight. The optical sight consists of a Zeiss-Diavari-Z 1.5-power x 6-mm zoom telescope. The weapon weighs 15.8 pounds with biped and 13.75 pounds without the biped. The NATO telescope mount allows almost any electro-optical or optical sight to be mounted to the weapon.

A-9. SPAIN

The 7.62-mm C-75 special forces rifle is the current sniper rifle of Spain. This weapon uses a manually operated Mauser bolt-action. It is equipped with iron sights and has telescope mounts machined into the receiver to allow for the mounting of most electro-optic or optic sights. The weapon weighs 8.14 pounds. An experienced firer can deliver effective fire out to 1,500 meters using Match ammunition.

A-10. SWITZERLAND

The Swiss use the 7.62-mm x 51 NATO SG 510-4SIG rifle with telescopic sight. The SG 510-4 is a delayed, blow-back-operated, 20-round, magazine-fed, semiautomatic or fully automatic weapon. With biped, telescope, and empty 20-round magazine, the weapon weighs 1229 pounds. It is 39.9 inches long with a 19.8-inch barrel and a muzzle velocity of 2,591 fps.

A-11. UNITED KINGDOM

The United Kingdom has four weapons designed for use by military snipers: the L42A1, Models 82 and 85 Parker-Hale, and L96A1.

a. **L42A1.** The L42A1 is a 7.62-mm x 51 single-shot, manually bolt-operated 10-round box magazine-fed conversion of the Enfield .303, Mark 4. It is 46.49 inches long with a barrel length of 27.51 inches. It comes equipped with metallic sights and 6-power x 42-mm LIAl telescope, and has a muzzle velocity of 2,748 fps.

b. **Model 82.** The Model 82 sniper rifle is a 7.62-mm x 51 single-shot, manually bolt-operated, 4-round internal magazine-fed rifle built upon a Mauser 98-action. It is equipped with metallic target sights or the more popular V2S 4-variable power x 10-mm telescope. It can deliver precision fire at all ranges out to 400 meters with a 99 percent chance of first-round accuracy. The weapon weighs 10.5 pounds and is 45.7 inches long. It is made of select wood stock and has a 25.9-inch, freefloated heavy barrel. An optional, adjustable biped is also available.

c. **Model 85.** The Model 85 sniper rifle is a 7.62-mm x 51 single-shot, manually bolt-operated, 10-round box magazine-fed rifle designed for extended use under adverse conditions. Its loaded weight of 30.25 pounds consists of an adjustable-for-length walnut stock with a rubber recoil pad and cold-forged, free-floated 27.5-inch heavy barrel. The popular telescope is 6-power x 44-mm with a ballistic cam graduated from 200 to 900 meters. This weapon is guaranteed first-round hit capability on targets up to 600 meters. It also provides an 85 percent first-round capability at ranges of 600 to 900 meters. Features include:

(1) An adjustable trigger.

(2) A silent safety catch.

(3) A threaded muzzle for a flash suppressor.

(4) A biped with lateral and swivel capabilities.

(5) An integral dovetail mount that accepts a variety of telescopes and electro-optical sights.

d. The L96A1 sniper rifle is a 7.62-mm x 51 single-shot, manually bolt-operated, 10-round box magazine-fed rifle weighing 13.64 pounds. It features an aluminum frame with a high-impact plastic, thumbhole-type stock, a free-floated barrel; and a lightweight-alloy, fully adjustable biped. The rifle is equipped with metallic sights that can deliver accurate fire out to 700 meters and can use the LIA1 telescope. The reported accuracy of this weapon is 0.75 MOA at 1,000 meters. One interesting feature of the stock design is a spring-loaded monopod concealed in the butt. Fully adjustable for elevation, the monopod serves the same purpose as the sand sock that the US Army uses.

A-12. UNITED STATES

The US Army sniper weapons are the M21 and M24 SWS. As with other countries, earlier production sniper rifles are still being used abroad to include the Ml, MIA-EZ and the M21. Other sniper weapon systems used by US forces are the USMC M40A1 and special application sniper rifles such as the RAI Model 500 and the Barrett Model 82.

a. **M40A1.** The M40A1 sniping rifle is a manually bolt-operated, 5-round internal magazine-fed 7.62-mm x 51 NATO weapon. This weapon is equipped with a Unertyl lo-power fixed telescope with a roil-dot reticle pattern as found in the M24's M3A telescope. The M40A1 is 43.97 inches long with a 24-inch barrel and weighs 14.45 pounds. It fires Ml 18 special ball ammunition and has a muzzle velocity of 2,547 fps and a maximum effective range of 800 meters.

b. **RAI Model 500.** The RAl Model 500 long-range rifle is a manually bolt-operated, single-shot weapon, and it is chambered for the caliber .50 Browning cartridge. Its 33-inch heavy, fluted, free-floating barrel, biped, and fully adjustable stock and cheekpiece weigh a total of 29.92 pounds. The weapon is equipped with a harmonic balancer that dampens barrel vibrations, a telescope with a ranging scope base, and a muzzle brake with flash suppressor. The USMC and USN use this weapon, which has a muzzle velocity of 2,912 fps.

c. **Barrett Model 82.** The Barrett Model 82 sniping rifle is a recoil-operated, 1 l-round detachable box, magazine-fed, semiautomatic weapon chambered for the caliber .50 Browning cartridge. Its 36.9-inch fluted barrel is equipped with a six-port muzzle brake that reduces recoil by 30 percent. It has an adjustable biped and can also be mounted on the M82 tripod or any mounting compatible with the M60 machine gun. This weapon has a pistol-grip-style stock, is 65.9 inches long, and weighs 32.9 pounds. The sighting system consists of a telescope, but no metallic sights are provided. The telescope mount may accommodate any telescope with l-inch rings. Muzzle velocity of the Model 82 is 2,849 fps.

A-13. THE FORMER RUSSIA

The Russians have a well-designed sniper weapon called the 7.62-mm Dragunov sniper rifle (SVD). The SVD is a semiautomatic, gas-operated, 10-round box, magazine-fed, 7.62-mm x 54 (rimmed) weapon. It is equipped with metallic sights and the PSO-1 4-power telescopic sight with a battery-powered, illuminated reticle. The PSO-1 also incorporates a metascope that can detect an infrared source. Used by the former Warsaw Pact armies, this thumbhole/ pistol-grip-style stocked weapon weighs 9.64 pounds with telescope and lo-round magazine. This weapon is 48.2 inches long with a 21.5-inch barrel, a muzzle velocity of 2,722 fps, and a maximum effective range of 600 to 800 meters.

A-14. THE FORMER YUGOSLAVIA

The former Yugoslav armed forces use the M76 semiautomatic sniping rifle. The M76 is a gas-operated, 10-round detachable box, magazine-fed, optically equipped 7.92-mm weapon. Variations of the weapon may be found in calibers 7.62-mm x 54 and 7.62-mm x 51 NATO. Believed to be based upon the FAZ family of automatic weapons, it features permanently affixed metallic sights, a pistol-grip-style wood stock, and a 4-power telescopic sight much the same as the Soviet PSO-1. It is graduated in NM-meter increments from 100 to 1,000 meters and has an optical sight mount that allows the mounting of passive nightsights. The M76 is 44.7 inches long with a 21.6-inch-long barrel. It weighs 11.2 pounds with the magazine and telescope, and it has a muzzle velocity of 2261 fps. A maximum effective range for the M76 is given as 800 meters with a maximum range of 1,000 meters.

APPENDIX B
M21 SNIPER WEAPON SYSTEM

The National Match M14 rifle (Figure B-1) and its scope makeup the M21 sniper weapon system. The M21 is accurized IAW United States Army Marksmanship Training Unit specifications and has the same basic design and operation as the standard M14 rifle (FM 23-8), except for specially selected and hand-fitted parts.

Section I
M21 SNIPER WEAPON SYSTEM

This section describes the general characteristics of the M21 SWS. The M21 has been replaced by the M24 (Chapter 2); however, the M21 is still in use throughout the US Army.

B-1. M21 DIFFERENCES
Significant differences exist between the M21 SWS and M24 SWS. These differences are as follows:

a. The barrel is gauged and selected to ensure correct specification tolerances. The bore is not chromium plated.

b. The stock is walnut and impregnated with an epoxy.

c. The receiver is individually custom fitted to the stock with a fiberglass compound.

d. The firing mechanism is reworked and polished to provide for a crisp hammer release. Trigger weight is between 4.5 to 4.75 pounds.

e. The suppressor is fitted and reamed to improve accuracy and eliminate any misalignment.

f. The gas cylinder and piston are reworked and polished to improve operation and to reduce carbon buildup.

g. The gas cylinder and lower band are permanently attached to each other.

h. Other parts are carefully selected, fitted, and assembled.

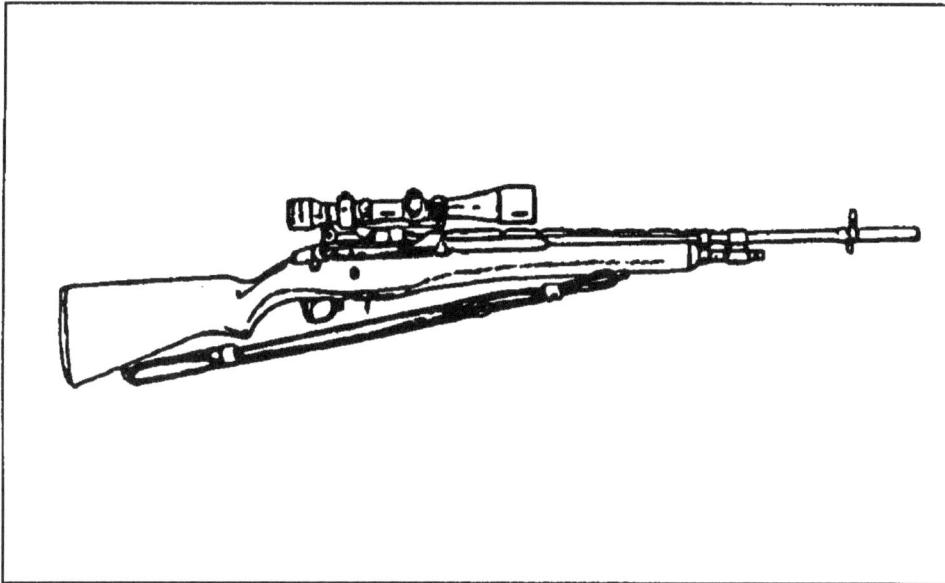

Figure B-1. National Match M14 rifle.

B-2. INSPECTION

If the sniper discovers a deficiency while inspecting the rifle, he reports it to the unit armorer. The following areas should be inspected:

a. Check the appearance and completeness of all parts. Shiny surfaces should be treated.

b. Check the flash suppressor for misalignment, burrs, or evidence of bullet tipping. The suppressor should be tight on the barrel.

c. Check the front sight to ensure that it is tight, that the blade is square, and that all edges and comers are sharp.

d. Check the gas cylinder to ensure it fits tightly on the barrel. The gas plug should be firmly tightened.

e. Check the forward band on the stock to ensure it does not bind against the gas cylinder front band.

f. Check the handguard. It should not bind against the receiver, the top of the stock, or the operating rod.

g. Check the firing mechanism to ensure the weapon does not fire with the safety "on," and that it has a smooth, crisp trigger pull when the safety is "off."

h. Check the rear sight tension by turning the aperture up to the "10" position. Then press down on top of the aperture with a thumb. If the aperture can be pushed down, the tension must be readjusted.

i. Check the stock for splits or cracks.

B-3. CARE AND MAINTANCE

Extreme care has been used in building the sniper rifle. A similar degree of attention must be devoted to its daily care and maintenance.

a. The rifle should not be disassembled by the sniper for normal cleaning and lubrication. Disassembly is performed only by the armorer during the scheduled inspections or repair. The armorer thoroughly cleans and lubricates the rifle at that time.

b. The following materials are required for cleaning the rifle

(1) Lubricating oil, general purpose (PL special).
(2) Lubricating oil, weapons (for below zero operation).
(3) Rifle bore cleaner.
(4) Rifle grease.
(5) Patches.
(6) Bore brush.
(7) Shaving brush.
(8) Toothbrush.
(9) Cleaning rags.

c. The recommended procedures for cleaning and lubricating the rifle are as follows:

(1) Wipe old oil, grease, and external dirt from the weapon.

(2) Clean the bore by placing the weapon upside down on a table or in a weapon cradle. hen, push a bore brush dipped in bore cleaner completely through the bore. Remove the bore brush and pull the rod out. Repeat this process four or five times.

(3) Clean the chamber (Figure B-2) and bolt face with bore cleaner and a chamber brush or toothbrush.

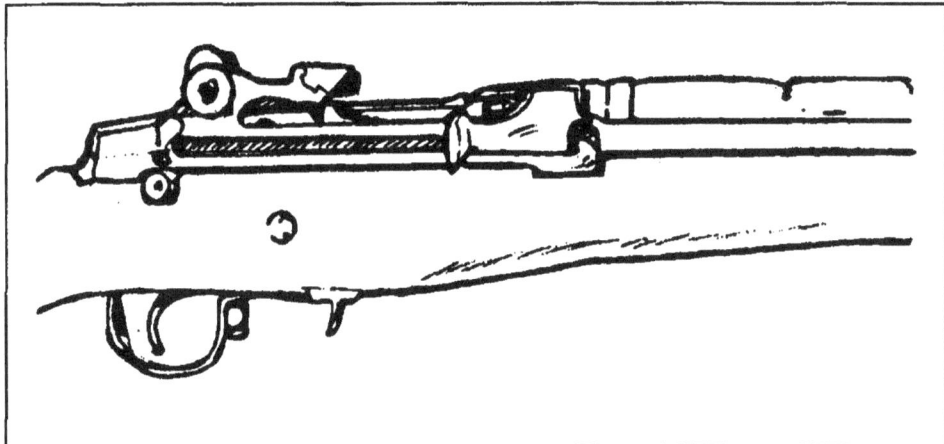

Figure B-2. M21 chamber.

(4) Clean the chamber, receiver, and other interior areas with patches dipped in RBC.

(5) Clean the bore by pulling clean patches through the bore until they come out of the bore clean.

(6) Wipe the chamber and interior surfaces with patches until clean.

(7) With the bolt and gas piston to the rear, place one drop of bore cleaner in between the rear band of the gas system and the lower side of the barrel. DO NOT PUT BORE CLEANER in the gas port. It will increase carbon buildup and restrict free movement of the gas piston.

(8) Lubricate the rifle by placing a light coat of grease on the operating rod handle track, caroming surfaces in the hump of the operating rod, the bolt's locking lug track, and in between the front band lip of the gas system and the metal band on the lower front of the stock.

(9) Place a light coat of PL special on all exterior metal parts.

B-4. LOADING AND UNLOADING

To load the M21, the sniper locks the bolt to the rear and places the weapon in the SAFE position. He inserts the magazine into the magazine well by pushing up, then pulling the bottom of the magazine to the rear until the magazine catch gives an audible click. To chamber a round, the sniper pulls the bolt slightly to the rear to release the bolt catch, then releases the bolt. To unload the M21, he locks the bolt to the rear and places the weapon in the SAFE position. Then he depresses the magazine release latch, and moves the magazine in a forward and downward motion at the same time.

B-5. REAR SIGHTS

The M21 sniper weapon system is equipped with National Match rear sights (Figure B-3). The pinion assembly adjusts the elevation of the aperture. By turning it clockwise, it raises the point of impact. By turning it counterclockwise, it lowers the point of impact. Each click of the pinion is 1 MOA (minute of angle). The hooded aperture is also adjustable and provides .5 MOA changes in elevation. Rotating the aperture so that the indication notch is at the top raises the point of impact .5 MOA. Rotating the indication notch to the bottom lowers the strike of the round. The windage dial adjusts the lateral movement of the rear sight. Turning the dial clockwise moves the point of impact to the right and turning it counterclockwise moves the point of impact to the left. Each click of windage is .5 MOA.

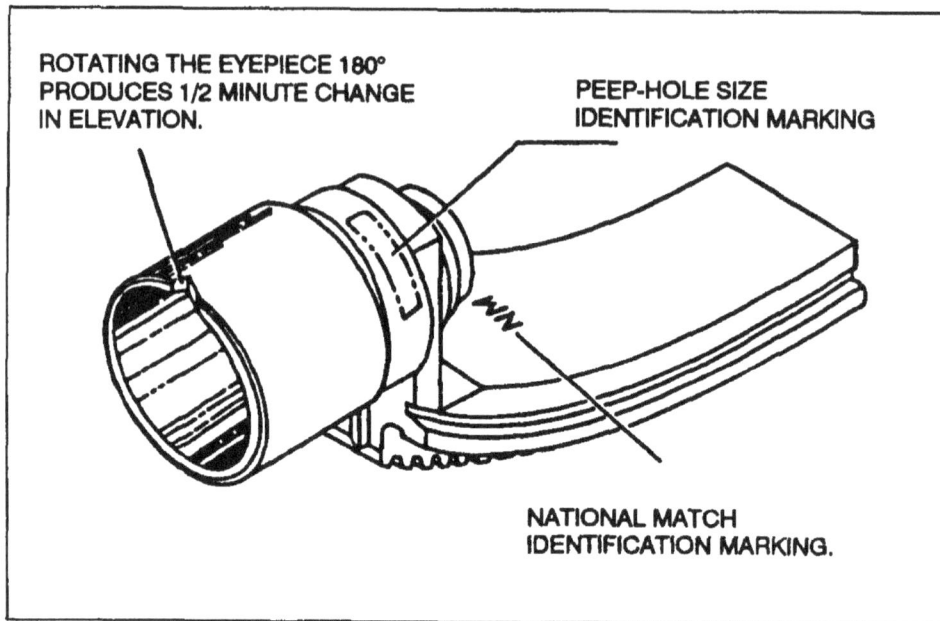

ROTATING THE EYEPIECE 180°
PRODUCES 1/2 MINUTE CHANGE
IN ELEVATION.

PEEP-HOLE SIZE
IDENTIFICATION MARKING

NATIONAL MATCH
IDENTIFICATION MARKING.

Figure B-3. National Match rear sight.

B-6. MALFUNCTIONS AND CORRECTIONS

Table B-1 contains pertinent information for the operator and serves as an aid to personnel who are responsible for restoring worn, damaged, or inoperative materiel to a satisfactory condition. If the weapon becomes unserviceable, it must be turned in for service by a school-trained National Match armorer.

MALFUNCTION	CAUSE	CORRECTION
FAILURE TO LOAD	DIRTY OR DEFORMED MAGAZINE	1. CLEAN OR REPLACE
	DAMAGED MAGAZINE TUBE	2. REPLACE MAGAZINE
	DIRTY MAGAZINE	3. CLEAN
	DAMAGED OR BROKEN MAGAZINE SPRING	4. REPLACE MAGAZINE
	DAMAGED OR BROKEN FOLLOWER	5. REPLACE MAGAZINE
	LOOSE OR DAMAGED FLOOR PLATE	6. REPLACE MAGAZINE

Table B-1. M21 malfunctions and corrections.

MALFUNCTION	CAUSE	CORRECTION
MAGAZINE INSERTS WITH DIFFICULTY	BENT OR DAMAGED MAGAZINE	7. REPLACE MAGAZINE
	EXCESSIVE DIRT IN RECEIVER	8. CLEAN
	ROUND NOT COMPLETELY SEATED IN MAGAZINE	9. REMOVE ROUND; INSERT PROPERLY
	DEFORMED OR OPERATING ROD SPRING GUIDE	10. EVACUATE TO AUTHORIZED ARMORER
	DAMAGED MAGAZINE LATCH	11. EVACUATE TO AUTHORIZED ARMORER
	MAGAZINE LATCH MOVEMENT RESTRICTED	12. CHECK MOVEMENT; CLEAN IF NECESSARY; IF BENT OR DISTORTED, EVACUATE TO AUTHORIZED ARMORER
MAGAZINE CANNOT BE RETAINED IN WEAPON	MAGAZINE LATCH DAMAGED	13. EVACUATE TO AUTHORIZED ARMORER.
	MAGAZINE LATCH SPRING DAMAGED	14. EVACUATE TO AUTHORIZED ARMORER
	MAGAZINE LATCH PLATE DAMAGED OR MISSING	15. REPLACE MAGAZINE
	DEFORMED OR DAMAGED OPERATING ROD SPRING GUIDE	16. EVACUATE TO AUTHORIZED ARMORER
	LOCKING RECESS AT TOP FRONT OF	17. REPLACE MAGAZINE
	MAGAZINE NOT FULLY INSTALLED	18. REMOVE; INSTALL CORRECTLY (MAKE SURE LATCH CLICKS)
FAILURE TO FEED	WEAK OR BROKEN SPRING	19. REPLACE MAGAZINE
	DAMAGED MAGAZINE	20. REPLACE MAGAZINE

Table B-1. M21 malfunctions and corrections (continued).

MALFUNCTION	CAUSE	CORRECTION
FAILURE TO FEED (CONTINUED)	DAMAGED OR DEFORMED STRIPPING LUG ON BOLT	21. EVACUATE TO AUTHORIZED ARMORER
	SHORT RECOIL	22. (SEE SHORT RECOIL)
	DIRTY AMMUNITION AND/OR MAGAZINE	23. CLEAN AMMUNITION AND/OR MAGAZINE
	WEAK OR BROKEN OPERATING ROD SPRING	24. EVACUATE TO AUTHORIZED ARMORER
	RESTRICTED MOVEMENT OF, OR DAMAGED OPERATING ROD25.	25. EVACUATE TO AUTHORIZED ARMORER
BOLT FAILS TO LOCK	CARTRIDGE CASE HOLDING BOLT OUT OF BATTERY	26. PULL BOLT CLOSE ASSEMBLY TO REAR AND REMOVE DEFORMED CARTRIDGE; CLEAN AMMUNITION AND/OR BARREL AND CHAMBER
	DIRTY CHAMBER	27. CLEAN CHAMBER
	EXTRACTOR DOES NOT SNAP OVER RIM OF CARTRIDGE CASE	28. EVACUATE TO AUTHORIZED ARMORER
	FROZEN OR BLOCKED EJECTOR SPRING AND PLUNGER	29. EVACUATE TO AUTHORIZED ARMORER
	RESTRICTED MOVEMENT OF, OR DAMAGED OPERATING ROD SPRING	30. EVACUATE TO AUTHORIZED ARMORER
	BOLT NOT FULLY ROTATED AND LOCKED IN RECEIVER	31. EVACUATE TO AUTHORIZED ARMORER
	WEAK OR BROKEN OPERATING ROD SPRING	32. EVACUATE TO AUTHORIZED ARMORER
	DAMAGED RECEIVER	33. EVACUATE TO AUTHORIZED ARMOR
FAILURE TO FIRE	BOLT NOT FULLY FORWARD AND LOCKED	34. (SEE BOLT FAILS TO LOCK)

Table B-1. M21 malfunctions and corrections (continued).

FM 23-10

MALFUNCTION	CAUSE	CORRECTION
FAILURE TO FIRE (CONTINUED)	DEFECTIVE AMMUNITION	35. REMOVE AMMUNITION
	FIRING PIN WORN, DAMAGED, OR MOVEMENT RESTRICTED	36. EVACUATE TO AUTHORIZED ARMORER
	BROKEN HAMMER	37. EVACUATE TO AUTHORIZED ARMORER
	WEAK OR BROKEN HAMMER SPRING	38. EVACUATE TO AUTHORIZED ARMORER
	HAMMER AND TRIGGER LUGS OR SEAR WORN OR BROKEN, CAUSING HAMMER TO RIDE THE BOLT ASSEMBLY FORWARD	39. EVACUATE TO AUTHORIZED ARMORER
SHORT RECOIL	GAS PLUG LOOSE OR MISSING	40. TIGHTEN PLUG IF LOOSE; EVACUATE TO AUTHORIZED ARMORER IF MISSING
	RESTRICTED MOVEMENT OF OPERATING ROD ASSEMBLY	41. EVACUATE TO AUTHORIZED ARMORER
	BOLT BINDING	42. CLEAN
	GAS CYLINDER LOCK NOT FULLY INSTALLED	43. EVACUATE TO AUTHORIZED ARMORER
	GAS PISTON RESTRICTED	44. EVACUATE TO AUTHORIZED ARMORER
	DAMAGED CONNECTOR ASSEMBLY	45. EVACUATE TO AUTHORIZED ARMORER
	PARTIALLY CLOSED SPINDLE VALVE	46. EVACUATE TO AUTHORIZED ARMORER
	DEFECTIVE AMMUNITION	47. REPLACE AMMUNITION
FAILURE TO EXTRACT	SPINDLE VALVE CLOSED	48. EVACUATE TO AUTHORIZED ARMORER
	CARTRIDGE SEIZED IN CHAMBER	49. REMOVE

Table B-1. M21 malfunctions and corrections (continued).

B-8

MALFUNCTION	CAUSE	CORRECTION
FAILURE TO EXTRACT (CONTINUED)	SHORT RECOIL	50. (SEE SHORT RECOIL)
	DAMAGED OR DEFORMED EXTRACTOR	51. EVACUATE TO AUTHORIZED ARMORER
	WEAK, DEFORMED, OR FROZEN EXTRACTOR PLUNGER ASSEMBLY	52. EVACUATE TO AUTHORIZED ARMORER
	RUPTURED OR SEPARATED CARTRIDGE	53. EVACUATE TO AUTHORIZED ARMORER
FAILURE TO EJECT	SHORT RECOIL	54. (SEE SHORT RECOIL)
	WEAK, DEFORMED OR FROZEN EXTRACTOR PLUNGER ASSEMBLY	55. EVACUATE TO AUTHORIZED ARMORER
FAIL TO HOLD BOLT REARWARD	DAMAGED OR DEFORMED MAGAZINE FOLLOWER	56. REPLACE MAGAZINE
	DAMAGED BOLT LOCK	57. EVACUATE TO AUTHORIZED ARMORER
	BOLT LOCK MOVEMENT RESTICTED	58. EVACUATE TO AUTHORIZED ARMORER
	WEAK OR BROKEN MAGAZINE SPRING	59. REPLACE MAGAZINE

Table B-1. M21 malfunctions and corrections (continued).

Section II
M21 SIGHTING DEVICES

A scope mounted on the rifle allows the sniper to detect and engage targets more effectively. The target's image in the scope is in focus with the aiming point (reticle). This allows for a more focused picture of the target and aiming point at the same time. Another advantage of the scope is its ability to magnify the target. This increases the resolution of the target's image, making it clearer and more defined. Keep in mind, a scope does not make a soldier a better sniper, it only helps him to see better.

B-7. AUTO-RANGING TELESCOPE

Auto-ranging telescopes are part of the M21 system. The two types of ARTs on the M21 system are the ART I and ART II. The basic design and operating principle of both scopes are the same. Therefore, they will be described together, but their differences will be pointed out.

B-8. ART I AND ART II SCOPES

The ART has a commercially procured 3- to 9-variable-power telescopic scopesight, modified for use with the sniper rifle. This scope has a modified reticle with a ballistic earn mounted to the power adjustment ring on the ART I (Figure B-4). The ART II (Figure B-5) has a separate ballistic cam and power ring. The ART is mounted on a spring-loaded base mount that is adapted to fit the M14. It is transported in a hard carrying case when not mounted to the rifle. The scopes on the M21 sniper weapon system can also be used for rough range estimation. Once the sniper team is familiar with the M21 and is accustomed to ranging out on targets, it makes a mental note of where the power adjust ring is set at various distances.

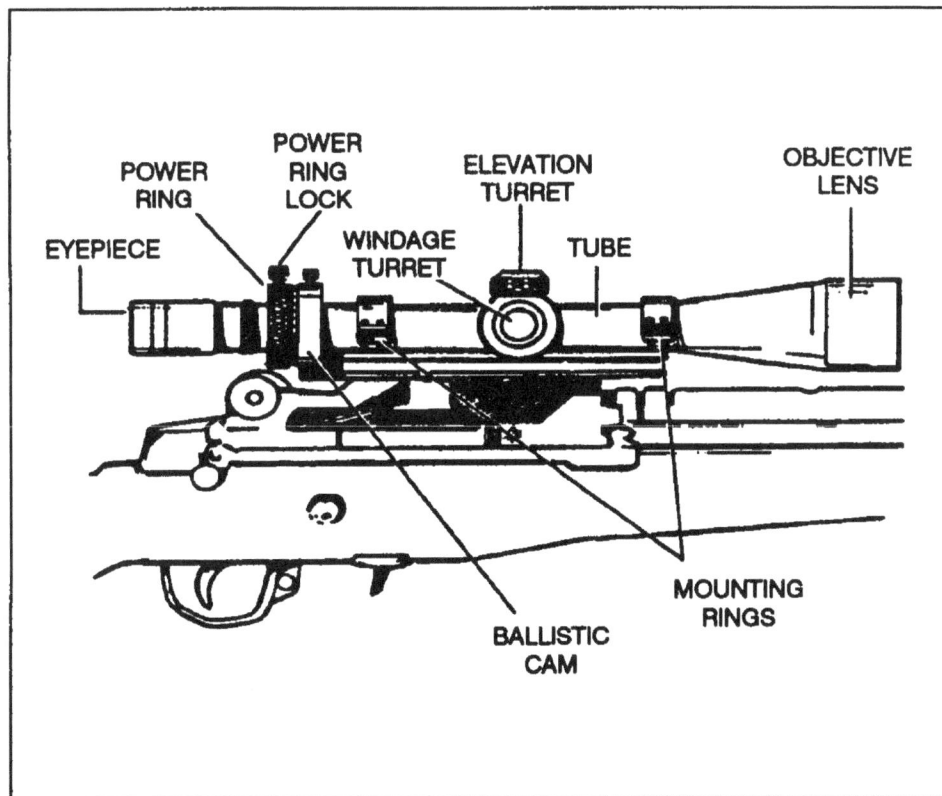

Figure B-4. ART I scope.

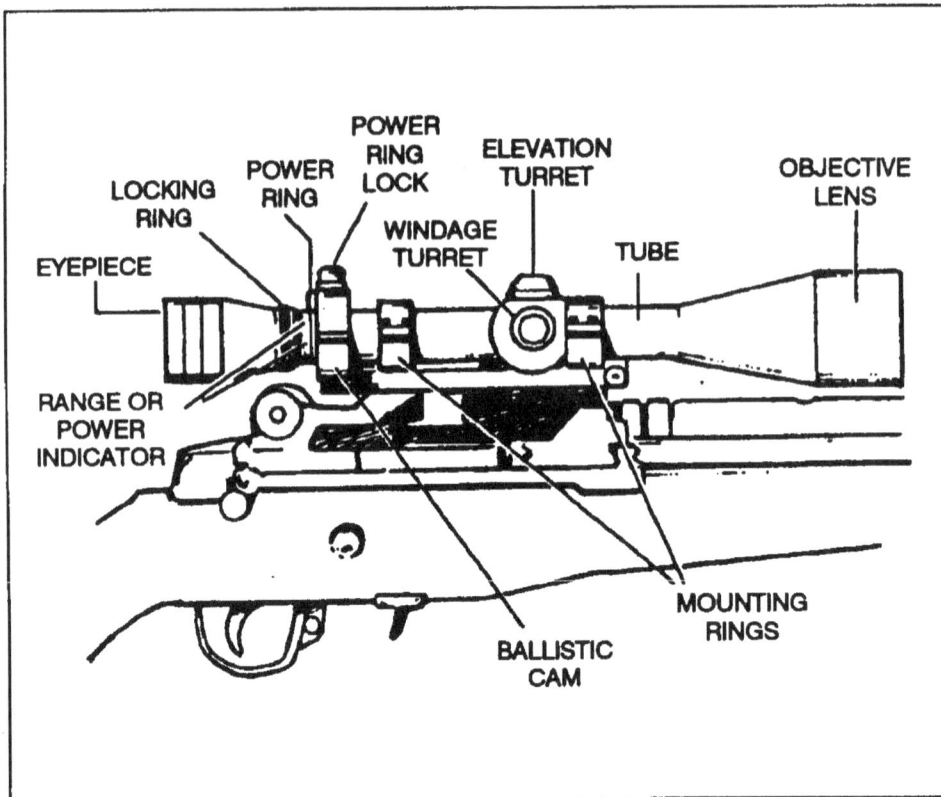

Figure B-5. ART II scope.

a. **Magnification.** The ART's increased magnification allows the sniper to seethe target clearly.

(1) The average, unaided human eye can distinguish detail of about 1 inch at 100 meters (1 MOA). Magnification, combined with well-designed optics, permit resolution of this 1 inch divided by the magnification. Thus, a 1/4 MOA of detail can be seen with a 4-power scope at 100 meters, or 1 inch of detail can be seen at 600 meters with a 6-power scope.

(2) The lens surfaces are coated with a hard film of magnesium fluoride for maximum light transmission.

(3) The elevation and windage turrets have dials on them that are located midway on the scope barrel and are used for zeroing adjustments. These dials are graduated in .5 MOA increments.

(4) These telescopes also have modified retitles The ART I scope has a basic cross hair design with two horizontal stadia lines that appear at target distances, 15 inches above and 15 inches below the horizontal

line of the reticle (Figure B-6). It also has two vertical stadia lines that appear at target distance, 30 inches to the left and 30 inches to the right of the vertical line of the reticle. The ART II scope reticle (Figure B-7) consists of three posts: two horizontal and one bottom vertical post. These posts represent 1 meter at the target's distance. The reticle has a basic cross hair with two dots on the horizontal line that appear at target distance, 30 inches to the left and 30 inches to the right of the vertical line.

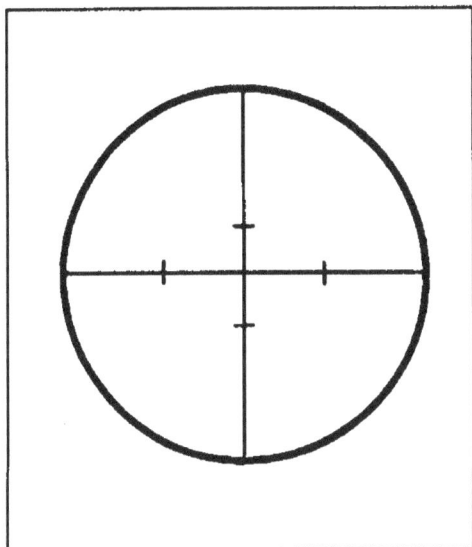

Figure B-6. ART I reticle.   **Figure B-7. ART II reticle.**

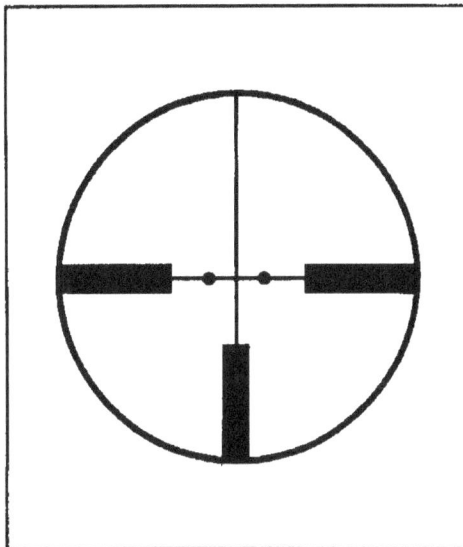

(5) A ballistic cam is attached to the power adjustment ring on the ART I scope. The ART II scope has a separate power ring and ballistic cam.

(6) The power ring on both scopes increases and decreases the magnification of the scope, while the ballistic cam raises and lowers the scope to compensate for elevation.

(7) Focus adjustments are made by screwing the eyepiece into or away from the scope barrel until the reticle is clear.

b. **Scope Mounts.** The ART mounts are made of lightweight aluminum consisting of a side-mounting plate and a spring-loaded base with scope mounting rings. The scope mount is designed for low-profile mounting of the scope to the rifle, using the mounting guide grooves and threaded hole(s) on the left side of the receiver. The ART I has one thumbscrew that screws into the left side of the receiver (Figure B-8). The ART II mount has two thumbscrews; one is screwed into the left side of the receiver, and the other is screwed into the cartridge clip guide in front of the rear sight (Figure B-9).

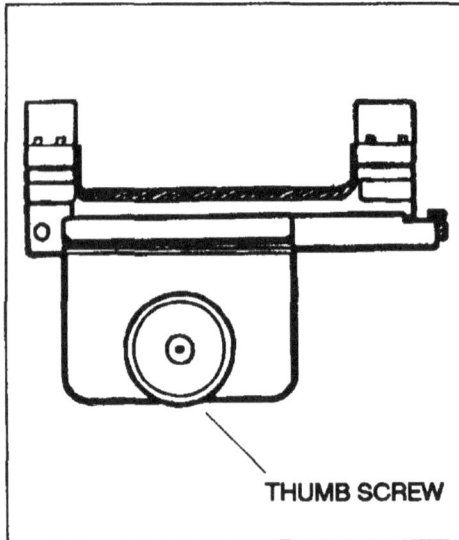

Figure B-8. ART I mount.

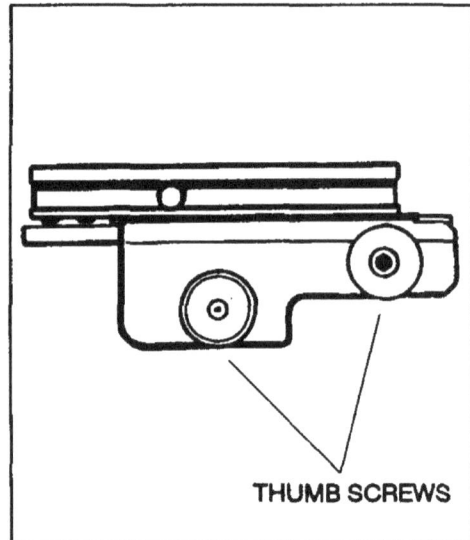

Figure B-9. ART II mount.

c. **Design and Operation.** The ART scopes are designed to automatically adjust for the needed elevation at ranges of 300 to 900 meters. This is done by increasing or decreasing the magnification of the scope until a portion of the target's image matches the represented measurement of the scope's reticle.

(1) For example, the power ring on the ART I scope can be adjusted until 30 inches of an object or a person's image (beltline to top of head) fits exactly in between the horizontal stadia lines (top stadia line touching top of the head and bottom stadia line on the beltline).

(2) Another example is to adjust the power ring on the ART II scope until 1 meter (about 40 inches) of a person's or an object's image appears equal to one of the posts in the reticle.

(3) When turning the power ring to adjust the target's image to the reticle, the sniper is also turning the ballistic cam. This raises or lowers the scope itself to compensate for elevation. Therefore, once the scope's magnification is properly adjusted in proportion to the target's image, the ballistic cam has at the same time adjusted the scope for the proper elevation needed to engage the target at that range.

(4) The ART II scope has a locking thumbscrew located on the power ring used for connecting and disconnecting the power ring from the ballistic cam. This allows the sniper to adjust the scope on target (auto-ranging mode), and then disengage the locking thumbscrew to increase magnification (manual mode) without affecting the elevation adjustment.

d. **Zeroing.** The ART scope should be zeroed at 300 meters. Ideally, this should be done on a known-distance range with bull's-eye-type targets. When zeroing the ART scope (Figure B-10), the sniper—

(1) Removes the elevation and windage caps from the scope.

(2) Turns the power adjustment ring to the lowest position. On the ART II scope, ensures that the locking thumbscrew is engaged and that the ballistic cam moves when the power ring is turned.

(3) Assumes a good prone-supported position that allows the natural point of aim to be centered on the target.

(4) Fires three rounds, using good marksmanship fundamentals with each shot.

(5) Makes the needed adjustments to the scope after placement of the rounds has been noted. He is sure to remember—

(a) That each mark on the elevation and windage dials equals .5 MOA (.5 MOA at 300 meters equals 1.5 inches.)

(b) That turning the elevation dial in the direction of the UP arrow will raise the point of impact; turning it the other direction will lower it.

(c) That turning the windage dial in the direction of the R arrow will move the point of impact to the right; turning it the other direction will move it to-the left.

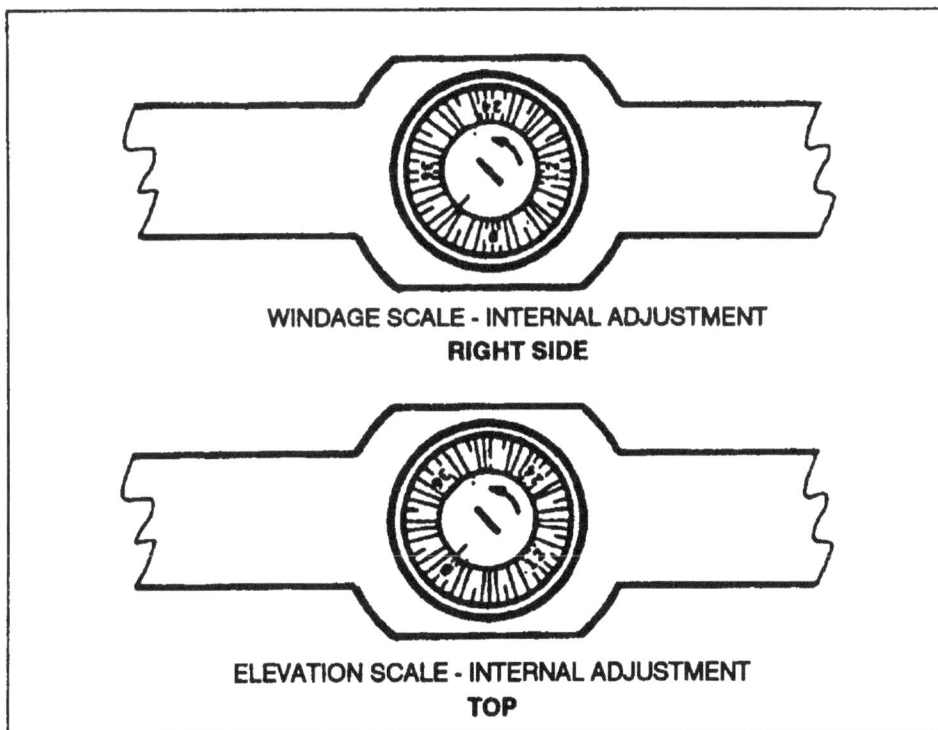

WINDAGE SCALE - INTERNAL ADJUSTMENT
RIGHT SIDE

ELEVATION SCALE - INTERNAL ADJUSTMENT
TOP

Figure B-10. Elevation and windage scales.

(6) Repeats the steps in paragraphs (4) and (5) above until two 3-round shot groups are centered on the target.

After the scope is properly zeroed, it will effectively range on targets out to 900 meters in the auto-ranging mode.

   e. **Zeroing and Calibrating of the M21 Iron Sights.**If the telescope is damaged, the sniper must use his backup sighting system-iron sights. Due to time constraints, it may be impossible or impractical to search through the data book to determine the needed elevation setting to engage a target at a specific range. Once the elevation dial has been calibrated to the sniper's individual zero for that particular rifle, targets can be engaged anywhere between 0 and 1,080 meters by using index lines.

   (1) The index lines on the elevation dial designate hundreds of yards to the target. Every other line is numbered with an even number, lines in between are the odd hundreds of yards-that is, the line marked with a number "2" is the 200-yard index line. The index line between the numbers 2 and 4 is the 300-yard index line. If the distance to the target is not in exact hundreds of yards, the elevation dial should be clicked between index lines to approximate the distance. If the target distance is less than 100 yards, the 100-yard setting should be used-the difference in impact is minimal.

   (2) To calibrate the elevation dial, the sniper must first zero the rifle at a known distance that correlates to one of the index lines on the elevation dial. (The recommended distance is 300 yards.) Once zeroing is completed, calibration involves the following steps:

   STEP 1: Turn the elevation dial forward (down, away from the sniper), and move the rear sight aperture assembly to its lowest setting (mechanical zero), counting the number of clicks. This number of clicks is elevation zero and must be remembered for use in the calibration process—for example, the number will be 10 clicks.

   STEP 2: Loosen the screw in the center of the elevation dial using a dime or screwdriver (about one turn) until the dial can be rotated forward Be careful not to loosen the screw too much or it may fall and become lost. It is critical that once the screw is loosened to never rotate the elevation dial clockwise (up or toward the sniper) during calibration. This could result in improper calibration.

   STEP 3: Turn the elevation dial forward (down, away from the sniper) until the index line on the receiver lines up with the index line on the dial that correlates to the distance at which the rifle was zeroed-for example, 300 yards. This is the index line between 2 and 4.

If the setting is passed (even by one click), rotate the elevation dial counterclockwise (down, away from the sniper) until the index lines match up. Never rotate the dial in the UP direction (clockwise, toward the sniper) with the screw in the elevation dial loose.

STEP 4: Remember the number of clicks (for example, 10) when zeroing the rifle and begin rotating the elevation dial counterclockwise (down, away from sniper). Count the clicks until the elevation dial has been rotated the same number of clicks that were on the rifle when zeroed. If too many clicks are used, start over at Step 3.

STEP 5: Now, hold the elevation dial, being careful not to allow it to rotate, then tighten the screw in the center of the elevation dial as tight as possible. Hold the elevation dial carefully with a pair of pliers to ensure the screw is tight.

STEP 6: To check the calibration, rotate the elevation dial to mechanical zero (all the way down), then count the number of clicks to zero. This should result in the index line on the receiver being lined up with the correct index line on the elevation dial (between 2 and 4). If this happens, the rear sight is now calibrated for elevation. If not, repeat Steps 1 through 5.

GLOSSARY

AARTY...... Army artillery

ADAM . . . . . . artillery delivered antipersonnel mine

aiming . . . . . . a marksmanship fundamental; refers to the precise alignment of the rifle sights with the target

ALICE . . . . . . all-purpose lightweight individual carrying equipment

AM . . . . amplitude modulation

antenna . . . . a device used to radiate or receive electromagnetic energy (usually RF)

antijamming . . a device, method, or system used to reduce or eliminate the effects of jamming

AP . . . . antipersonnel

APFT . . . . . . Army Physical Fitness Test

APICM . . . . antipersonnel improved conventional munition

armorer . . . . one who services and makes repairs on small arms and performs similar duties to keep small arms ready for use

ARNG . . . . . . .Army National Guard

ART . . . . . . .auto-ranging telescope

ARTEP . . . . . . .Army Training and Evaluation Program

AVLB . . . . . . armored vehicle launched bridge

AWADS . . . . adverse weather aerial delivery system

ball . . . . the projectile the bullet

ballistics . . . . a science that deals with the motion and flight characteristics of projectiles

BDU . . . . . . battle dress uniform

BMNT . . . . . . beginning morning nautical twilight

breath control a marksmanship fundamental refers to the control of breathing to help keep the rifle steady during firing

bullet drop . . how far the bullet drops from the line of departure to the point of impact

bull's-eye target . any target with a round black circle and scoring rings Normally used in competitive marksmanship training

butt plate . . . . . metal or rubber covering of the end of the stock on the rifle

CALFEX . . . . . combined arms live-fire exercise

cartridge . . . . . a complete round of ammunition

CAB . . . . close air support

CLGP . . . . cannon-launched guided projectile

CAP . . . . cleaner, lubricant, preservative

cm . . . . centimeter

CMF . . . . career management field

counterpoise . . . a conductor or system of conductors used as a substitute for a ground in an antenna system

CP . . . . concrete-piercing

CQ . . . . charge of quarters

crack and thump . a method to determine the general direction and distance to an enemy firer who is shooting at you

cradle . . . . a vise-like mechanism that holds a weapon in a secured position during test firing

CS . . . . a chemical agent (tear gas)

CW . . . . continuous wave

dia . . . . diameter

dipole . . . . a radio antenna consisting of two horizontal rods in line with each other with their ends slightly separated

DPICM . . . . . . dual-purpose improved conventional munition

DIG . . . . date-time group

DZ . . . . drop zone

E&E . . . . evasion and escape

ECM . . . . electronic countermeasures

EDGE . . . . emergency deployment readiness exercise

EEL . . . . essential elements of information

EVENT. . . . . . . end of evening nautical twilight

effective wind . . the average of all the varying winds encountered

electromagnetic wave . . .a wave propagating **as** a periodic disturbance of the electromagnetic field and having a frequency in the electromagnetic spectrum

elevation adjustment . . . rotating the front sight post to cause the bullet to strike the higher or lower on the target

EMP . . . . electromagnetic pulse

EPW . . . . enemy prisoner of war

ERP . . . . end-route rally point

eye relief . . . . . the distance from the firing eye to the rear sight; eye relief is a function of stock weld

F . . . . Fahrenheit

FDC . . . . . . fire direction center

FFL . . . . . . final firing line

FFP . . . . . . final firing position

FLOT . . . . . . forward line of own troops

FM . . . . frequency modulated

FO . . . . forward observer

fps . . . . feet per second

FRAGO . . . . fragmentary order

freq . . . . frequency

FSK . . . . . . frequency-shift keying

ft . . . . feet

FTX . . . . . . field training exercise

ground . . . . . . a metallic connection with the earth to establish ground (or earth) potential

HAHO . . . . . . high altitude, high opening

half-wave antenna . . . . . an antenna whose electrical length is half the wavelength of the transmitted or received frequency

HALO . . . . . . high altitude, low opening

HC . . . . hydrogen chloride

HE . . . . high explosive

HF . . . . high frequency

hrs . . . . hours
Hz . . . . hertz

IAW . . . . in accordance with
ID . . . . identification
illum . . . . illumination
in . . . . inches
insulator . . . . . a device or material that has a high electrical resistance
interference . . . any undesired signal that tends to interfere with the desired signal

IRP . . . . initial rally point

jamming . . . . deliberate interference intended to prevent reception of signals in a specific frequency band

KIM . . . . keep-in-memory (exercise game)

laser . . . . light amplification by simulated emission of radiation
LBE . . . . loading-bearing equipment
LAX . . . . live-fire exercise
line of departure . the line defined by the bore of the rifle or the path the bullet would take without gravity
line of sight . . . . a straight line from the eye through the aiming device to the point of aim
L/R.. . . . . left/right
LR . . . . laser range finder
LSA . . . . lubricating oil, weapons, semifluid
LZ . . . . landing zone

m . . . . meters
MEDEVAC . . . . medical evacuation
METT-T . . . . . mission, enemy, terrain, troops and time available
MHz . . . . megahertz

midrange trajectory/maximum ordinate . . the highest point the bullet reaches on its way to the target this point must be known to engage a target that requires firing underneath an overhead obstacle, such as a bridge or a tree; inattention to midrange trajectory may cause the sniper to hit the obstacle instead of the target

MIJI . . . . meaconing, intrusion, jamming, and interference

MILES . . . . . . multiple-integrated laser engagement system

min . . . . minute(s)

mm . . . . millimeter

MOA . . . . an angle that would cover 1 inch at a distance of 100 yards, 2 inches at 200 yards, and so on; each click of sight adjustment is equal to one minute of angle

MOPP . . . . . . mission-oriented protection posture

MOUT . . . . . . military operations on urbanized terrain

mph . . . . miles per hour

MRE . . . . meal, ready-to-eat

MTP . . . . mission training plan

muzzle velocity . the speed of the bullet as it leaves the rifle barrel, measured in feet per second; it varies according to various factors, such as ammunition type and lot number, temperature, and humidity

NATO . . . . . . North Atlantic Treaty Organization

NBC . . . . nuclear, biological, chemical

NCO . . . . noncommissiotted officer

NGF . . . . naval gunfire

NOD . . . . night observation device

NSN . . . . national stock number

OIR . . . . other intelligence requirements

OP . . . . observation post

OPORD . . . . . operation order

OPSEC . . . . . operations security

optical sight . . sight with lenses, prisms, or mirrors used in lieu of iron sights

ORP . . . . objective rally point

O-T . . . . observer-target

PD . . . . point-detonating
PFC . . . . private first class
P I R . . . . priority intelligence requirements
POC . . . . point of contact
point of aim . . . the exact spot on a target the rifle sights are
                 aligned with
point of impact . . the point that a bullet strikes; usually considered in
                 relation to point of aim
PSG . . . . platoon sergeant
PT . . . . physical training
PW . . . . prisoner of war
PZ . . . . pickup zone

QRF . . . . quick-reaction force
quarter-wave antenna . . an antenna with an electrical length that is
                 equal to one-quarter wavelength of the signal being
                 transmitted or received

RMMS . . . . . . remote antiarmor mine system
range card . . . . small chart on which ranges and directions to various
                 targets and other important points in the area under
                 fire are recorded
RAP . . . . rocket-assisted projectile
RBC . . . . . rifle bore cleaner
recoil . . . . the rearward motion or kick of a gun upon firing
RECONREP . . . reconnaissance report
retained velocity . the speed of the bullet when it reaches the target
                 due to drag, the velocity will be reduced
RF . . . . radio frequency
RFA . . . . restrictive fire area
RFL . . . . restrictive fire line
round . . . . may refer to a complete cartridge or to the bullet
RSTA . . . . reconnaissance, surveillance, and target acquisition

S1 . . . . adjutant
S2 . . . . intelligence officer
S3 . . . . operations and training officer

S4 . . . . .supply officer

SALUTE . . . . . size, activity, location, unit, time, and equipment

SAW . . . . squad automatic weapon

SEO . . . . sniper employment officer

servomechanism . . . . . .an automatic device for controlling large amounts of power by using small amounts of power

SFC . . . . sergeant first class

SGT . . . . sergeant

SHELREP . . . . shelling report

shot group . . . . a number of shots fired using the same aiming point which accounts for rifle, ammunition, and firer varibility three shots are enough, but any number of rounds may be fired in a group

sight alignment . placing the center tip of the front sight post in the exact center of the rear aperature

silhouette target . a target that represents the outline of a man

single sideband . a system of radio communications in which the carrier and either the upper or lower sideband is removed from AM transmission to reduce the channel width and improve the signal-to-noise ratio

SIR . . . . specific information requirements

SITREP . . . . . situation report

SM . . . . smoke munitions

SOI . . . . signal operation instructions

SOP . . . . standing operating procedure

SP . . . . self-propelled

SPC . . . . specialist

SPIES . . . . . . special patrol insertion/ extraction system

SPOTREP . . . . spot report

SSB . . . . single sideband

STAB . . . . . . a system for extracting personnel by helicopter

STANAG . . . . . Standardization Agreement

static . . . . sharp, short bursts of noise on a radio receiver caused by electrical disturbances in the atmosphere or by electrical machinery

steady position . . the first marksmanship fundamental, which refers to the establishment of a position that allows the weapon to be held still while it is being fire

stock weld . . . . the contact of the cheek with the stock of the weapon

STRAC . . . . . . standards in training commission

STX . . . . situational training exercise

**supported
position** . . . . . . any position that uses something other than the body to steady the weapon (artificial support)

S W S . . . . sniper weapon system

TAB . . . . tactical air command

TFFP . . . . tentative final firing position

time of flight . . . the amount of time it takes for the bullet to reach the target from the time the round exits the rifle

TOC . . . . tactical operations center

TOW . . . . tube-launched, optically tracked wire-guided (missile)

trajectory . . . . . the path of the bullet as it travels to the target

TRC . . . . training readiness condition

TAP . . . . target reference point

unidirectional . . in one direction only

unsupported position . . . any position that requires the firer to hold the weapon steady using only his body (bone support)

USAF . . . . United States Air Force

USAR . . . . United States Army Reserve

U S C . . . .. United States Code

USMC . . . . . . United States Marine Corp

USN . . . . United States Navy

VHF . . . . very high frequency

VT . . . . variable time

wavelength . . . . the distance a wave travels during one complete cycle; it is equal to the velocity divided by the frequency

windage adjustment . . . moving the rear sight aperture to cause the bullet to strike left or right on the target

WP. . . . . white phosphorus

zeroing . . . . . . adjusting the rifle sights so bullets hit the aiming point at a given range

REFERENCES

SOURCES USED

These are the sources quoted or paraphrased in this publication.

*STANAG 2020. Operational Situation Reports. 13 February 1986

*STANAG 2022. Intelligence Reports. 29 September 1988.

*STANAG 2084. Handling and Reporting of Captured Enemy Equipment and Documents. 26 June 1986.

*STANAG 2096. Reporting Engineer Information in the Field. 13 Jul 1988.

*STANAG 2103. Reporting Nuclear Detonations, Radioactive Fallout, and Biological and Chemical Attacks, and Predicting Associated Hazards. 12 July 1988.

*STANAG 2934. Artillery Procedures—AARTY-1. 26 November 1990.

*STANAG 3204. Aeromedical Evacuation.

STANAG 6004. Meaconing, Intrusion, Jamming, and Interference Report. 20 March 1984.

DOCUMENTS NEEDED

These documents must be available to the intended users of this publication.

ARTEP 7-92-MTP. Infantry Scout Platoon/ Squad and Sniper Team. 16 March 1989.

DA Form 5785-R. Sniper's Data Card. June 1989.

DA Form 5786-R. Sniper's Observation Log. June 1989.

DA Form 5787-R. Sniper's Range Card. June 1989.

DA Form 5788-R. Military Sketch. June 1989.

DA Form 7325-R. Concealment Exercise, July 1994.

DA Form 7326-R. Concealed Movement Exercise Scorecard, July 1994.

---

*This source was also used to develop this publication.

DA Form 7327-R. Target Detection Exercise Scorecard, July 1994.

DA Form 7328-R. Range Estimation Exercise Scorecard, July 1994.

DA Form 7329-R. Qualification Table No. 1 Scorecard, July 1994.

DA Form 7330-R. Qualification Table No. 2 Scorecard, July 1994.

*DA Pam 350-38. Training Standards in Weapon Training. 1990.

*TM 9-1005-306-10. Operator's Manual for 7.62 mm M24 Sniper Weapon System (SWS). 23 June 1989.

*TM 9-1265-211-10. Operator's Manual for Multiple Integrated Laser Engagement System (MILES)---Simulator System, Firing Laser, M89... Squad Automatic Weapon (SAW). 28 February 1989.

READINGS RECOMMENDED

These reading contain relevant supplemental information.

*FM 5-20. Camouflage. 20 May 1968.

*FM 5-36. Route Reconnaissance and Classification. 10 May 1985.

*FM 6-30. Observed Fire Procedures. 17 June 1985.

FM 6-121. Field Artillery Target Acquisition. 13 December 1984.

*FM 7-93. Long-Range Surveillance Unit Operations. 9 June 1987.

FM 8-10-4. Medical Platoon Leader's Handbook. 16 November 1990

FM 8-35. Evacuation of the Sick and Wounded. 22 December 1983.

*FM 17-98-1. Scout Leader's Handbook. 24 September 1990.

FM 21-26. Map Reading and Navigation. 5 July 1993.

FM 21-75. Combat Skills of the Soldier. 3 August 1984.

FM 23-8. M14 and M14A1 Rifles and Rifle Marksmanship. 15 April 1974.

FM 23-9. M16A1 and M16A2 Rifle Marksmanship. 3 July 1989.

FM 23-31. 40-mm Grenade Launchers M203 and M79. (To Be Published.)

*FM 24-1. Signal Support in the AirLand Battle. 15 October 1990.

*FM 24-18. Tactical Single-Channel Radio Communications Techniques. 30 September 1987.

*This source was also used to develop this publication.

FM 34-3. Intelligence Analysis. 15 March 1990.

FM 90-3(HTF). Desert Operations (How to Fight). 19 August 1977.

FM 90-4. Air Assault Operations. 16 March 1987.

FM 90-5(HTF). Jungle Operations (How to Fight). 16 August 1982.

FM 90-6. Mountain Operations. 30 June 1980.

FM 90-10(HTF). Military Operations on Urbanized Terrain (MOUT) (How to Fight). 15 August 1979.

TC 31-24. Special Forces Air Operations. 9 September 1988.

TC 31-25. Special Forces Waterborne Operations. 30 October 1988.

TM 11-666. Antennas for Radio Propagation. 9 February 1953.

TM 11-5855-262-10-1. Litton Model M972/ M973. 15 June 1987.

TM 11-5860-201-10. Operator's Manual: Laser Infrared Observation Set, AN/ GVS-5. 2 February 1982.

INDEX

SNIPER'S DATA CARD

For use of this form, see FM 23-10; the proponent agency is TRADOC

DISTANCE TO TARGET _____ METERS

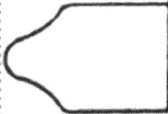

RANGE:	RIFLE AND SCOPE NO		DATE	ELEVATION		WINDAGE	
				USED	CORRECT	USED	CORRECT
AMMO	LIGHT	MIRAGE	TEMP	HOUR			

LIGHT

WIND

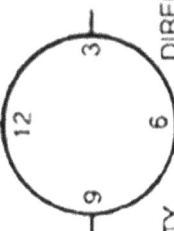

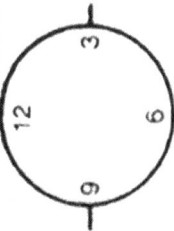

VELOCITY   DIRECTION

SHOT	1	2	3	4	5	6	7	8	9	10	REMARKS
ELEV.											
WIND											

C											
A											
L											
L											

HOLD

NOTE: THE REQUIRED TARGETS WILL BE DRAWN IN BY HAND TO MEET THE NEEDS OF THE UNIT.

DA FORM 5788-R, JUN 89

SNIPER'S OBSERVATION LOG

For use of this form, see FM 23-10; the proponent agency is TRADOC

SHEET _____ OF _____ SHEETS

ORIGINATOR:

DATE/TIME:

LOCATION:

SERIAL	TIME	GRID COORDINATE	EVENT	ACTIONS OR REMARKS

DA FORM 5785-R, JUN 89

SNIPER'S RANGE CARD

For use of this form, see FM 23-10; the proponent agency is TRADOC

NAME:

METHOD OF OBTAINING RANGE

RANGE		
ELEVATION		
WINDAGE		

TEMPERATURE		WIND		
HIGH	LOW	VELOCITY	DIRECTION	

TRP 1			TRP 2			TRP 3		
AZIMUTH	DISTANCE		AZIMUTH	DISTANCE		AZIMUTH	DISTANCE	
DESCRIPTION			DESCRIPTION			DESCRIPTION		

DA FORM 5786-R, JUN 89

MILITARY SKETCH

For use of this form, see FM 23-10;

REMARKS:

REMARKS:

SKETCH NAME: _____

GRID COORDINATE: _____

WEATHER: _____

= ←

MAGNETIC

AZIMUTH

SKETCH # _____

OF _____

SCALE _____

NAME:

RANK:

DATE/TIME:

DA FORM 5787-R, JUN 89

CONCEALMENT EXERCISE SCORECARD

Exercise Number ____

For use of this form, see FM 23-10; the proponent agency is TRADOC

DATA REQUIRED BY PRIVACY ACT OF 1974.

AUTHORITY: 10 USC 3012(g)/Executive Order 1974. PRINCIPAL PURPOSE(S): Evaluates individual training. ROUTINE USE(S): Evaluates individual proficiency. SSN is used for positive identification purposes only. MANDATORY OR VOLUNTARY DISCLOSURE AND EFFECT ON INDIVIDUAL NOT PROVIDING INFORMATION: Voluntary. Individuals not providing information cannot be rated/scored on a mass basis.

Last name	First	MI	Rank	SSN	Unit

Date	Weather\visibility			Score	

	Points	Deducted	Total
If the sniper			
• Was detected without the aid of optics (first 2 minutes)	2	0	2
• Was detected with the aid of optics (18 minutes)	1	0	3
• Was detected when assistant trainer was within 10 feet of sniper	1	0	4
• Properly identied the number within 30 seconds	1	0	5
• Failed to properly identify the number	0	3	2
• Fired first shot, not detected	4	0	6
• Fired second shot, not detected	1	0	7
• Maintained stable firing position (support)	2	0	9
• Properly adjusted weapon's scope for range and windage	1	0	10

(Check one of the target indicators.)

☐ Contrast to background
☐ Muzzle blast
☐ Muzzle flash
☐ Improper movement techniques
☐ Improper camouflage
☐ Shine
☐ Outline
☐ Sound

NOTES:

1. If the sniper was caught trying to identify the number, score 4 points.

2. If muzzle blast/flash is detected, deduct 1 point from total score.

3. Failing to comply with training standards and objectives (such as unnecessary movement, premature fire, outside prescribed boundries) will result in termination of the exercise and a score of zero.

_____
Trainer's signature

_____
Sniper's signature

DA FORM 7325-R, JUL 94

CONCEALED MOVEMENT EXERCISE SCORECARD

Exercise Number_____

For use of this form, see FM 23-10; the proponent agency is TRADOC

DATA REQUIRED BY PRIVACY ACT OF 1974.

AUTHORITY: 10 USC 3012(g)/Executive Order 9397. PRINCIPAL PURPOSE(S): Evaluates individual training. ROUTINE USE(S): Evaluates individual proficiency. SSN is used for positive identification purposes only. MANDATORY OR VOLUNTARY DISCLOSURE AND EFFECT ON INDIVIDUAL NOT PROVIDING INFORMATION: Voluntary. Individuals not providing information cannot be rated/scored on a mass basis.

Last name	First	MI	Rank	SSN	Unit

Date	Weather\visibility			Score	

	Points	Deducted	Total
If the sniper			
• Was detected moving to FFL	.0	.0	.0
• Was detected moving in FFL	.6	.0	.6
• Fired first round shot, not detected . . .	.2	.0	.8
• Was not detected when assistant trainer is within 10 feet of sniper	.2	.0	.10
• Properly identified number (within 30 seconds)	.2	.0	.12
• Failed to properly identify number . . .	.2	.0	.14
• Was not detected when assistant trainer is within 5 feet of sniper	.2	.0	.16
• Fired second shot, not detected	.2	.0	.18
• Maintained stable firing position (support) . . .	.1	.0	.19
• Properly adjusted weapon's scope for range and windage . . .	.1	.0	.20

(Check one of the target indicators.)

☐ Contrast to back-ground
☐ Muzzle blast
☐ Muzzle flash
☐ Improper movement techniques

☐ Improper camouflage
☐ Shine
☐ Outline
☐ Sound

NOTES:

1. If muzzle blast/flash is detected, deduct 1 point from total score.

2. Failing to comply with training standards and objectives (such as unnecessary movement, premature fire, outside prescribed boundaries) will result in termination of the exercise and a score of zero.

REMARKS: Explain in detail on back the reason for sniper's detection.

_____   _____
Trainer's signature   Sniper's signature

DA FORM 7326-R, JUL 94

	A	B	C	D	E	F	G	H	
7									7
6									6
5									5
4									4
3									3
2									2
1									1
0									0

SKETCH NAME: _____

GRID COORDINATE: _____

WEATHER: _____

= _____
Magnetic Azimuth

Sketch # ____
of ____
Block Scale:
_____

SKETCH NAME: _____

GRID COORDINATE: _____

DATE: _____ TIME: _____

#	SIZE	SHAPE	COLOR	CONDITION	APPEARS TO BE	GRID BOX LOC.
1						
2						
3						
4						
5						
6						
7						
8						
9						
10						

RANGE ESTIMATION EXERCISE SCORECARD

Exercise Number _____

For use of this form, see FM 23-10; the proponent agency is TRADOC

DATA REQUIRED BY PRIVACY ACT OF 1974. PRINCIPAL PURPOSE(S): Evaluates individual proficiency. ROUTINE USE(S): Evaluates individual training. SSN is used for positive identification purposes only. MANDATORY OR VOLUNTARY DISCLOSURE AND EFFECT ON INDIVIDUAL NOT PROVIDING INFORMATION: Voluntary. Individuals not providing information cannot be rated/scored on a mass basis.

Last name	First	MI	Rank	SSN	Unit

Date	Weather\visibility		Score		

EYE ESTIMATION +- 15%	BINOCULAR ESTIMATION +- 10%	M3A TELESCOPE ESTIMATION +- 5%
1 _____	1 _____	1 _____
2 _____	2 _____	2 _____
3 _____	3 _____	3 _____
4 _____	4 _____	4 _____
5 _____	5 _____	5 _____
6 _____	6 _____	6 _____
7 _____	7 _____	7 _____
8 _____	8 _____	8 _____
9 _____	9 _____	9 _____
10 _____	10 _____	10 _____

1. Within three minutes, the range to the target is estimated at each point, using the naked eye, binoculars, and the M3A telescope. Estimations must be performed in the order listed.

2. Once an estimate is recorded, it cannot be changed; it will be counted as incorrect. However, the M3A telescope estimate may be changed before the next set of estimates are recorded.

3. The use of calculators is encouraged.

4. This is an individual exercise. Any sniper that talks or tries to look at another sniper's scorecard is terminated from the exercise.

5. If there are any questions, the trainer will assist you.

_____
Trainer's signature

_____
Sniper's signature

DA FORM 7328-R, JUL 94

MAKE A SKETCH OF THE SECTOR YOU ARE ASSIGNED TO COVER.

	A	B	C	D	E
1					
2					
3					
4					
5					

QUALIFICATION TABLE No. 1 SCORECARD
Exercise Number_____

For use of this form, see FM 23-10; the proponent agency is TRADOC

(circle one)
Record or practice

Last name	First	MI	Rank	SSN	Unit

Date	Weather\visibility		Score

TARGET (meters)	1st Round	2d Round	Miss	TARGET (meters)	1st Round	2d Round	Miss
200	_____	_____	_____	500	_____	_____	_____
300	_____	_____	_____	400	_____	_____	_____
325	_____	_____	_____	325	_____	_____	_____
375	_____	_____	_____	400	_____	_____	_____
500	_____	_____	_____	600	_____	_____	_____
600	_____	_____	_____	500	_____	_____	_____
500	_____	_____	_____	700	_____	_____	_____
375	_____	_____	_____	325	_____	_____	_____
600	_____	_____	_____	300	_____	_____	_____
700	_____	_____	_____	200	_____	_____	_____

_____ x10 _____ x5=_____

Trainer's signature   Sniper's signature

DA FORM 7329-R, JUL 94

QUALIFICATION TABLE No. 2 SCORECARD
Exercise Number____

For use of this form, see FM 23-10; the proponent agency is TRADOC

DATA REQUIRED BY PRIVACY ACT OF 1974.

AUTHORITY: 10 USC 3012(g)/Executive Order 9397. PRINCIPAL PURPOSE(S): Evaluates individual training. ROUTINE USE(S): Evaluates individual proficiency. SSN is used for positive identification purposes only. MANDATORY OR VOLUNTARY DISCLOSURE AND EFFECT ON INDIVIDUAL NOT PROVIDING INFORMATION: Voluntary. Individuals not providing information cannot be rated/scored on a mass basis.

(circle one)
Record or practice

Last name	First	MI	Rank	SSN	Unit

Date	Weather\visibility	Score

TARGET (meters)	1st Round	2d Round	Miss	TARGET (meters)	1st Round	2d Round	Miss
200	_____	_____	_____	900	_____	_____	_____
300	_____	_____	_____	850	_____	_____	_____
325	_____	_____	_____	800	_____	_____	_____
375	_____	_____	_____	750	_____	_____	_____
600	_____	_____	_____	700	_____	_____	_____
500	_____	_____	_____	900	_____	_____	_____
600	_____	_____	_____	500	_____	_____	_____
700	_____	_____	_____	400	_____	_____	_____
750	_____	_____	_____	325	_____	_____	_____
800	_____	_____	_____	300	_____	_____	_____
850	_____	_____	_____				

_____x10 _____x5=_____

_____
Trainer's signature

_____
Sniper's signature

DA FORM 7330-R, JUL 94

By Order of the Secretary of the Army:

GORDON R. SULLIVAN
General, United States Army
Chief of Staff

Official:

MILTON H. HAMILTON
Administrative Assistant to the
Secretary of the Army
06908

DISTRIBUTION:

Active Army, USAR, and ARNG:
To be distributed in accordance with DA
Form 12-11E, requirements for FM 23-10,
Sniper Training (Qty rqr block no. 1335)

*U.S. Government Printing Office: 1994—528-027/80156

PIN: 072777-000

www.ingramcontent.com/pod-product-compliance
Lightning Source LLC
Chambersburg PA
CBHW081144270326
41930CB00014B/3035